Suffolk County Council
Libraries & Heritage

Natural
Plant Dyes

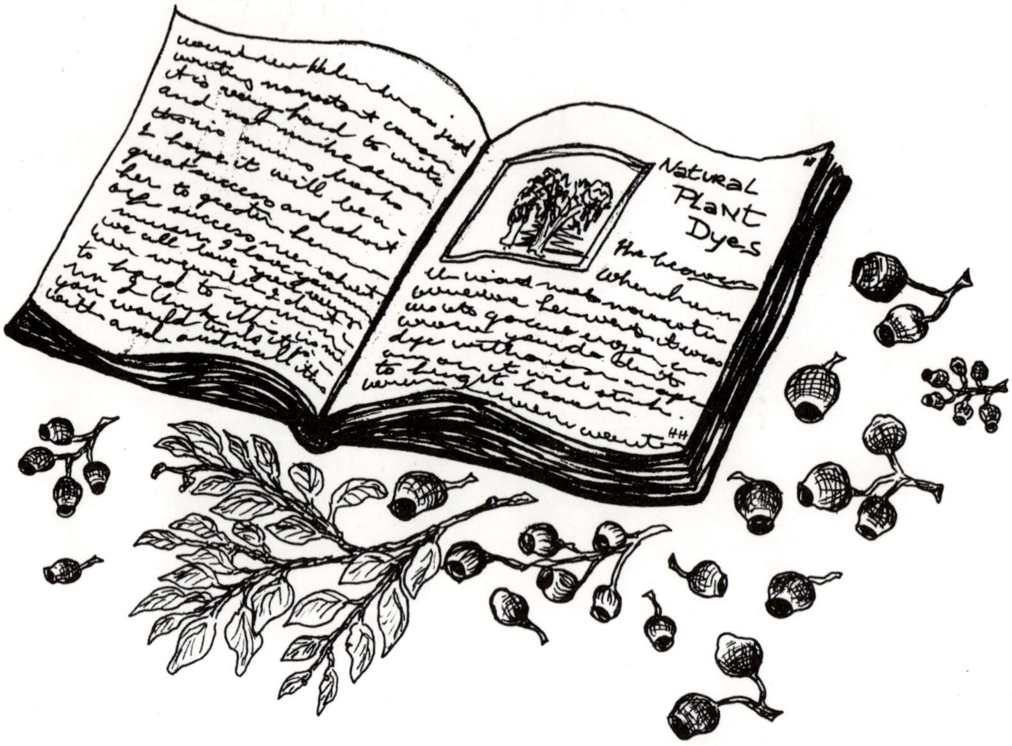
Natural Plant Dyes

Natural Plant Dyes

Judith V. Hallett

Kangaroo Press

Acknowledgments

I acknowledge with gratitude the encouragement given me by my dear friend Jean Carman whose work with dyes has so inspired me. I would also like to thank Pauline Ballinger who was Secretary of the Queensland Spinners, Weavers and Dyers Group during my time of office on the Board, and always brought out the best in me.

I also want to thank my children, Vivianne, Paul and Helen, for their encouragement and help. Vivianne and Paul made valuable suggestions, and Helen did most of the artwork. I also appreciate the efforts of my husband Mel who typed the manuscript. I would particularly like to thank Joan Apthorp for her photography of the samples and closeups. The balance of the photographs were a joint effort between Mel and myself.

Thanks also to members of the Queensland Spinners, Weavers and Dyers Group for plant specimens given to me over the years, and for their interest in my experiments. Thanks also to the Botany Section of the Department of Primary Industries in Brisbane for help with the botanical names of some of the plants. Also sincere thanks to special friends who believe I am worthy to contribute something to this craft.

© Judith V. Hallett

First published in 1992
Reprinted in 1993
Second edition published in 1996 by Kangaroo Press Pty Ltd
3 Whitehall Road Kenthurst NSW 2156 Australia
(PO Box 6125 Dural Delivery Centre NSW 2158)
Typeset by G. T. Typesetters Pty Limited
Printed in Hong Kong by Colorcraft Ltd

ISBN 0 86417 752 6

Contents

Introduction 6
Ancient Dyes 7
Let's Collect 8
Experimenting 9
Wool preparation 10
Are you afraid to dye? 12
Be prepared to dye 13
Dye pots 14
Amounts of dye matter (ratios) 14
Water 15
Time your dyeing 15
Where would you like to dye? 16
Mordanting 19
Mordant recipes 21
Come dye with me 24
Dyeing with cochineal 26
Dyeing with lichen 31

Come dye in the forest 32
Cotton on to dyeing 30
Dyeing for silk 36
Dyeing for children 38
Are you dyeing to know? 41
Making a dye sample book 42
Uses for dye samples 44
Storing dyestuffs 44
I'm dyeing to tell you about... 45
Lists 47
  1. Colour list 48
  2. Plants for colours list 54
  3. Botanical to common names list 60
  4. Wood shavings for colours list 62
References 63
Index 64

Introduction

This book is a practical and simple guide, a textbook for dyemaking with plants. It is written in simple terms, using easy methods. Even if you are a novice you will find dyeing exciting to try, and if you have tried dyeing before, it will inspire you to experiment further. The reader must understand that the natural dyeing process is very simple, not a long involved process at all. Quite a number of colours can be obtained in a few hours work. Some years ago, for example, I conducted a workshop for a group of thirteen students at The Australian Woolshed in Brisbane. In one 6-hour workshop we ended up with 112 different coloured samples of wool, using twenty-five different plants—and we still took time off for lunch!

Australian plants are the source of an incredible range of colours. Mixing certain colours gives even more shades. Listings at the back of the book cover well over one hundred different plants which have been tested; the colours produced with and without mordants have all been noted. Use these lists as a basis for your own experiments.

Reading the book through first before you begin your experiments will help you get a feel for the process. Experiment with the unknown—use a new plant and see what wonderful new colours you find.

Natural wool, which is most consistently receptive to dye, and is easily obtainable throughout Australia, has been used for all the tests in this book. Craft shops and spinners' and weavers' groups always stock fleece and some spun wool.

Let your imagination go and find God's hidden colours in this beautiful world. Go forth and dye!

Ancient dyes

Making dyes from plants is an ancient art. Way back in history we read of clothing being dyed red, blue, purple and other shades. People living in the Mediterranean region used shellfish of the genus *Murex* to dye their clothing purple—the famous colour known as Tyrian purple. Others found that the stains from various berries were useful as dyes.

Red, a popular colour for soldier's uniforms from early times, came from the root of a plant called madder *(Rubia tinctoria)*. Another source of red dye is the scale insect *Dactylopius coccus*, found on cactus plants. Much of this dye, known as cochineal, came from Mexico.

Blue came from a number of plants of the genus *Indigofera*, especially *Indigofera tinctoria*. A lot of preparation is required to extract this dye and to make it useable on fibres.

All dyed fibres will fade in time but most plant-dyed pieces, provided they are well looked after, remain beautiful. The colours become muted but are still a joy to behold.

The old tapestries of past centuries were all woven with yarns dyed with plant dyes. These tapestries are still beautiful; they may be faded but they remain real treasures.

Animal fibres such as wool, mohair, silk, dog hair, camel hair, rabbit fur, and so on, dye very well. Plant fibres such as cotton and flax need extra preparation. It is wise to investigate fully the ways of preparing these fibres before you dye. The same dye plants used on plant fibres will often give colours different to the ones they produce on animal fibres.

Note: At the time of writing (1992), some of the ancient dyes mentioned could be obtained from Commission Dyers Pty Ltd, 7 Pinn Street, St Marys, South Australia 5042.

Let's collect

Dyestuff is best collected in dry weather. Leaves and flowers seem to absorb extra water in rainy periods and the colours obtained then are not as good. Collecting dye materials can be fun, and educational, too. You will need a lot of large bags and a pen for identifying the contents of each bag. (Remember to observe the rules for protection of native plants.)

A working knowledge of the plants of your district would be a help, but even if you can only identify a few you can still collect. Take care with known poisonous plants. Reading through a few gardening books beforehand will help you to get to know your plants. A field guide to the plants in your area would be invaluable, especially if you need to check something along the way. I have found three books very helpful, and have often used them—*A Field Guide to Australian Trees* by Ivan Holliday and Ron Hill, *A Gardener's Guide to Eucalypts* by Ivan Holliday and Geoffrey Watton, and *A Field Guide to Weeds in Australia* by Charles Lamp and Frank Collet. These books are easy to use because they have photographs of the whole plant and either a closeup or detailed sketch of the plant's parts. A number of other field guides are also worth investing in. A list of reference books is given on page 63.

There are thousands of different plants; even if you spent a lifetime collecting and experimenting, you would still not cover the full range.

If you are collecting leaves, always collect mature leaves as these have the greatest amount of dye. With flowers, always use those in full bloom or flowers that have reached maturity and are ready to drop their petals. Berries must be fully ripe. When it comes to bark, older bark is the best to use.

Oldest is best isn't always the Golden Rule, but most times it is so. One exception is bracken fern—I have found the young fronds give the best shade of green. If you have the patience and the spare wool—try some young shoots and buds of the material you have collected, and compare the results with older material.

Don't overlook the weeds growing in your garden or by the roadside in favour of the established plants in your garden. Lots of the plants we list as weeds are quite valuable to the home dyer. Wildflowers and grasses yield dyes, too; so the world of dye plants extends beyond your own garden.

Dyestuffs can be used fresh or dried and used later. Hang bunches of leaves in an airy place to dry, then place them in bags for future use. Always note on the bag the common plant name, the botanical name if you can establish it, the season collected and the place. Different seasons produce slightly different intensities of colour, but they are usually in the same tones. Dried plants are often hard to identify, so label as you go.

Experimenting

Always make notes as you dye, listing the amount of dye stuff, the amount of water, the time taken to extract the dye, the time taken to get the desired colour, and the different colours obtained with various mordants. I find mordanting is necessary to obtain the full colour range from each plant. (Mordanting is explained later, on page 19.) A dye experiment sheet, like this one, is a handy way of summarising results.

Dye experiment sheet

Done by ....

Project dyed...

Dye plant..

Common name ....

Botanical name ....

Leaves, bark or flowers ....

Green or dried ....

Date of collection ....

Mordant ....

Type of dye pot ....

Amount of water ....

Type of water (rain, bore, town) ....

Method ....

Date ....

Quantity used ....

....

Wool type used ....

Comments ....

Samples ....

Colour test: Only small amount required: 30 g leaves to 625 ml water.

....

....

....

Wool preparation

You don't necessarily have to use homespun wool for dyeing with plant dyes. Commercial pure wool or wool and nylon will also dye very successfully. (Acrylic fibres do not dye well. They come out looking quite insipid; the results are usually very disappointing.)

Weigh the wool on your kitchen scales before you wet it, so you can calculate quantities of dye, water and mordant more easily.

Soak homespun wool in a tub of cold water overnight. (See the sketch.) This loosens the dirt and softens the grease. Then wash in warm water with a product especially recommended for woollen garments. Treat the wool gently; don't rub it. Repeat the warm wash only if the wool is still dirty. Rinse the wool first in warm water, then cold, before proceeding to mordanting and dyeing.

If you are using commercial wool, there is no need to wash it in soap. Just rinse it in warm water and proceed to the mordanting and dyeing.

Always, and I stress always, rinse wool in very warm to hot water before putting it into the dye pot. Damp wool accepts the dye more readily. Rinsing the wool before dyeing removes any excess mordant, which might affect the quality of the dye colour.

Shrinkage

Sometimes people ask about shrinkage of the wool. Pure wool does shrink a small amount when simmered for a long time (two hours, say). Always allow for a little more length than you think you'll need for a project to cover this problem. If the dye takes on the fibre very quickly, that is, in the first fifteen minutes, you will find only minimal shrinkage occurs.

I have found that I generally lose only one or perhaps two centimetres per metre—and that is only with homespun Tukiedale, simmered for two hours. Commercial pure wool shrinks less than that, and Merino only a little also. Remember, less time in the pot, less shrinkage, so take the wool out when the colour is to your satisfaction.

Another question often asked is whether the simmering, mordanting or dyeing of wool changes the feel of it in any way. The answer is no. Even simmering for up to two hours does not change the feel of the wool.

Mordanting can affect the fibres *only* if too much mordant is used. If you follow exactly the recipes for mordanting given on pages 21–23, no problems should arise.

Simmering the wool in the dyeing process and rinsing afterwards in hot to warm, then cool water is the right way to go. Turn the fibre over gently with the stir stick and keep submerged. Remember *don't boil*, only simmer; *don't rinse in cold water* (use hot to warm, then cool water); and *don't stir violently*. Remember these three things and you will never have matted wool.

The only problem I have come across in the eleven years I have been dyeing was that once some Corriedale wool had a tendency to cling together and look matted. As the wool was in a skein the strands were easily parted. That particular yarn is today knitted up into a beautiful shawl.

Whenever possible dry the wool yarn or fleece outside in a good breeze. The result will be soft, fluffy yarn which is a delight to work with.

One must remember, however, that if the wool is a coarse-fibre wool, like Tukiedale or maybe Border Leicester, both used for rug making, it will stay coarse. A wool like Merino or Polwarth or Romney, which starts out soft and fluffy, will stay that way. The way it starts out is the way it will end up if you are gentle with it.

Soaking and cleaning yarn or fibre

Method	Ingredients		Time
SOAK	yarn in water		overnight
REMOVE	yarn and pour out water		
MIX	soap and hand-hot water		
ADD	yarn or fibre		
WASH	yarn in soap and hand-hot water	gently all together	until clean
REMOVE	yarn and pour out water		
RINSE	yarn in hot water		until rinse water runs clear
PROCEED	to premordanting or dyeing or		
STORE	wet until next day	or dry	for future use

Are you afraid to dye?

I certainly hope not. Plant dyeing is simple.

Don't be afraid to experiment. All plants, even the most unlikely candidates, yield dye to a greater or lesser extent.

Try a simple experiment. Get an old saucepan, enamel or stainless steel for preference, but aluminium will do.

Now the dye plants. The plants in this list only need boiling up, the wool to be added and then simmered for half an hour or so. No mordant is used.

> Young bracken fern gives a pretty green
> Aluminium plant gives a soft tan
> Celery leaves give a lemon-green
> Brown onion skins give golds/rust/ginger
> Eucalypts (most varieties) give fawn/brown to deep rust
> Red camellia flowers give fawn/grey
> Nasturtium leaves give pale lemon-green
> Chrysanthemum gives pale lemon
> Red rose petals give a soft brown
> Loquat leaves give a pinky fawn

Try some of these. Fill the pot with chopped-up celery leaves, for example, and cover it with water. Boil up the celery leaves for about half an hour, then let it cool a little off the stove. Take some homespun wool, or commercial white wool, either pure wool or wool and nylon. Make up a small skein, tying a few loose loops around the skein to keep it together.

Wet the wool with warm to hot water, put it in the dye pot, and simmer for half an hour. Check from time to time to see how it is taking the colour. Take out the plant matter if you wish. Dyeing with eucalyptus leaves takes a little longer; you can simmer the wool for up to two hours without harming it.

Using a pot made of non-reactive material such as enamel or stainless steel gives clearer colours, but an old aluminium pot will do for experimenting. (Non-reactive materials don't give off any metal residue.)

Rinse the wool in warm to hot water, then cool water, and dry it. If you are happy with the results, next time you can try something more interesting, by mordanting the wool before dyeing it. Pages 24–25 give you a more detailed procedure for dyeing.

Mordants

Most mordants, which are used to fix colours, are metal salts that work by causing the wool fibres to open up and accept the dye more readily. By using wool treated with four or five different mordants you can get four or five different colours out of the one dye pot, as the mordants react with the dye to produce sometimes quite surprising results.

The mordants I use are alum, iron, copper, chrome and tin. Only very small quantities of each of these are needed.

One project, one pot

It's always best to dye all the yarn needed for one project in the one dye pot. Later dye lots will always be very slightly different in colour. If it isn't possible to dye all the yarn together, owing to the size of the pot or the size of the project, divide the yarn into even lots and dye. Later, when using the dyed yarn in the project, use a small amount from each lot in turn. In this way any slight unevenness of colour will blend less noticeably.

Be prepared to dye

Dyeing with natural plant dyes is very rewarding. It is very simple and cheap because you can use what is growing in your own garden. Every plant is a dye source in itself. Some colours are subtle, some are strong. The shades always blend, never clash.

Most plants yield their colour just by boiling up—that is, cover them with water and heat. Sometimes a different dye can be obtained by using the ammonia method. This involves soaking the leaves or flowers in one part ammonia (clear or cloudy) to two parts water in a sealed jar. Let the jar stand for a few days to a week and see if another colour is released.

Using the two methods of dye release, and the various mordants, you can obtain many shades of colour from each plant. If you need a blend of colours for a project, whether it be a knitted or crocheted garment, a weaving or a tapestry, all the colours you need can be obtained reasonably quickly.

Dye plants

Every plant will release a dye. Some colours are very strong, others are insignificant. Some apparently disappointing colours can still be valuable; an unexciting colour can be incorporated with other, brighter colours in a garment or weaving to enhance the brighter colours. Never discard a disappointing result—somewhere in the future you are bound to need that colour for something.

Your own garden has a wealth of colour even if you only have a few trees or shrubs. The different parts of a plant can provide different coloured dyes. Some interesting variations can result from boiling up leaves, flowers, bark, seeds, nuts or pods, and roots in separate pots. Experimenting in this way can result in a range of shades of colour from one plant. Sometimes only one plant will provide all the colours for a project.

You can get at least five different colours from the leaves of just one plant. Using the flowers, another five, totally different, colours may be obtained. Berries or nuts will give another five colours. I have found that the colours from a plant's leaves blend beautifully with the colours from the flowers. Roots are generally used only where the plant is an annual and will therefore be dug up, or where a tree has blown over during a storm so that the roots are accessible.

Always collect specimens for dyeing in a responsible way, and never completely strip a plant of leaves or flowers. 'Selective pruning' is the watchword. Think about the plant, and be a conservationist. Most plants respond well to a bit of pruning. Always observe the laws relating to collection of native flora, especially in National Parks and Forestry areas. Permits can be obtained for specimen collecting.

Don't collect during wet weather as plants absorb extra moisture during this time. Plants collected during dry periods will show brighter and stronger colours—you will be much happier with the results.

Don't overlook the importance of herbs, vegetable plants, fruit trees and fruit. The smell of herbs simmering in a pot can be marvellous wafting through the house. Some green leafy vegetables release a good dye, while some fleshy fruits and vegetables release very little.

Test a small piece first before cutting up a whole supply. Squash a berry between your fingers to check if a dye colour is present. If it stains your fingers it's generally worthwhile using. Dye from berries, however, tends to fade quite a bit as time goes on. Some are better eaten than used for dyeing. Sometimes the water content of fruits and vegetables is so high that you need much less water than usual. A high water content is often an indication that the plant (part) is not very useful for dyeing.

Dye pots

The best and clearest colours are obtained using non-reactive pots. By non-reactive I mean enamel, stainless steel or Pyrex pots, but they need not be the expensive types. Stainless steel heats up more quickly, and so saves on gas or electricity. Enamel is good, but it must be unchipped. Disposal stores usually have cheap pots large enough for dyeing a kilogram of yarn. Secondhand shops are sometimes an excellent source.

Always keep your dyepots for dyeing. Never use them for food because of the risk of poisoning from traces of plant material or mordants, some of which are very dangerous.

Amounts of dye matter (ratios)

As a general rule, use *at least* 400 g of dye material to 100 g of wool, more if you like. If you want a strong colour use a 10 to 1 ratio, that is, 1000 g (1 kg) of leaves or whatever to 100 g wool. For mid shades I use a 5 to 1 ratio, 500 g leaves to 100 g wool. For pale shades equal quantities of leaves and wool generally works. Don't add extra water to 'dilute' a colour—dyeing doesn't work that way.

It all depends on what colour you want for a set project. Experimenting with small quantities of wool and strong dye first, will tell you what range of colours can be obtained from a particular plant. Paler shades will always come when you exhaust the dye pot.

To exhaust the dye pot, simmer a small skein of prepared wool for twenty minutes. Remove the wool and rinse. Repeat the process with another skein, Remove the wool and rinse. Keep repeating this process until the wool stays white. It may take a while but you will get a terrific range of colour, from deep to pastel shades.

It's easier to control the dyeing process if you have a good strong pot of dye. Do remember to weigh the plant matter and note the information on a dye experiment sheet, because you may want to repeat the colour sometime.

Passiflora

Water

The type of water used in the dyeing process can affect the end result. A town water supply has added chemicals; rainwater is generally free of chemicals. Bore water will frequently alter colours because of its high level of natural chemicals. I use rainwater where I can; I feel I get the best, clearest colours this way.

Please don't be put off from trying dyeing if you haven't got rainwater. You will still get very pleasant results and it's often interesting to compare your colours with colours from other areas where the water is different. Remember—there are no hard and fast rules with dyeing as long as you enjoy learning and are pleased with the finished result. *Always* keep records of what you dye and the water used, along with the area and the season, and you will find you can repeat these colours again and again.

Time your dyeing

Sometimes it takes longer to extract dye out of plants. Test a small amount of wool in the pot if necessary.

Timing is crucial and must be recorded to obtain the same colour again. Check the guide to dye extraction times below.

The ammonia and water method of dye extraction is described on page 27. Some plants don't release dye with water, but using ammonia as well will release the dye. Sometimes the process may take some days before you see any results.

Guide to dye extraction times

Plant part	Boiling time	Preparation
Berries	15-20 minutes	Crush with potato masher
Flowers	20 minutes	
Leaves	30-60 minutes or longer	Stiff leaves must be boiled longer—presoak overnight
Bark	60 minutes or longer	Presoak at least 24 hours
Pinecones	60 minutes or longer	Presoak at least 24 hours
Roots	30-60 minutes	Presoak 12-24 hours

Where would you like to dye?

In the house? In the garden? In the shed? In the park?

Dyeing can be done in the kitchen; some plants, however, are much better dealt with outside, especially those that release a strong odour—lots of fresh air is necessary. Only a few plants are really smelly, but one of the added advantages of being in the garden is that unwanted vegetable matter from the dye pot can be emptied directly into the compost heap. Another advantage is the fact that spillages won't matter so much, while a third is that additional plants can be collected while simmering the dye pot and chopping up extra material can be done without worrying too much about the mess. Anyway, it's always pleasant working in the fresh air.

Even though onion skins, marigolds and a few others give off strong or unpleasant odours during dye extraction, they *do not* impart the smell to the yarn. Once the yarn has dried, no odour is present.

Turn up the heat

What type of heating should you use if you're going to work outside in a courtyard or other sheltered corner? You can use an open fire in the barbecue, but it's very difficult to regulate the heat. I find the gas barbecue or a portable electric hotplate much more useful for this reason. With the electric hotplate, work within a short distance of a power point, as using an extension lead is inadvisable.

Dyeing can be loads of fun if a few friends work together, each simmering a pot of different dye. At the end of the day everyone has fun comparing the end results.

Above left: *Ardisia crispa* The red berries of this shrub, which grows to about 1 m, are used to produce the dye. The colours obtained can range from light grey to steel grey with touches of gold or brown

Above: *Bougainvillea* x 'Brittania' This is a rambling creeper with brilliantly coloured papery flowers useful for producing a green dye which gives colours of pale green, citrus and olive green

Left: *Bixa orellana* This shrub grows to about 2 m, producing pink flowers which in turn produce prickly seed pods. The seeds yield an orange coloured dye, which dyes fibres shades of soft gold, orange and soft brown

Above: Celery *(Apium graveolens)* The dried or fresh leaves of this useful culinary vegetable are useful also in dyemaking, the dye from them producing lovely shades of lime to olive green

Left: The bright yellow flowers of *Cassia corymbosa* will dye fibres dull gold to olive gold. The leaves are also useful, producing a dye that colours yarn a pale green through to citrus green

17

Above: Argyle apple *(Eucalyptus cinerea)* This eucalypt, which grows to about 10 m and has quite a few common names, gives a very strong dye. Its silver-blue foliage is used in dried flower arrangements. As a dye source it is very valuable. The colour range runs from pale apricot to rust red and every possible gold/rust shade in between. It needs no mordant

Above: Beret made of handspun wool, handknitted. The red centre was dyed with *Eucalyptus cinerea*

Left: Detail of wool wall hanging dyed with *Eucalyptus cinerea*. Both items made by Karlene Lewis

Mordanting

Don't let the rather grim sound of the word 'mordanting' put you off the process. You will get better, more interesting results if you mordant your yarn, and many more colours.

Mordanting is really very simple, although you must follow the safety procedures outlined on page 23. It's merely a matter of adding a chemical (alum, chrome, copper, iron or tin) to boiling water, mixing, adding more water, then the wool, and simmering (see the diagrams on page 20).

The five most commonly used mordants are alum (potassium aluminium sulphate), chrome (potassium dichromate), copper (copper sulphate), iron (ferrous sulphate) and tin (stannous chloride). Their recipes appear on pages 21-22.

I use these mordants to treat the wool before dyeing as I find I get better results in the end. I generally use tin only as an additive at the end of dyeing, to get all the dye possible out of the dye pot. A small pinch of tin will give another shade of colour, but if you like, you can mordant with tin and then dye.

The skeins of wool should be left loose through mordanting and dyeing, not twisted, and tied loosely at intervals around the skein. This prevents the wool becoming tangled and knotted.

Because four or five skeins of wool, each treated with a different mordant, can be dyed simultaneously for four or five different results, it is a very wise move to establish a marking system to identify each skein of yarn. The knotting system of marking, explained below, helps identify the skein after dyeing, making it possible to get the same result in the future.

My system, simplicity itself, is this: after making up the skeins to the size I want, I tie the ends together with a knot, leaving about 8 cm of free yarn. On these ends I make the knots: no knot for unmordanted yarn, 1 knot for yarn mordanted with alum, 2 knots for chrome, 3 knots for copper, 4 knots for iron, and 5 knots for tin. It's easy to remember because the mordants are in alphabetical order.

Why do we mordant?

A mordant is a chemical, sold in the form of powder or crystals, which when mixed with water and heated opens up the scales on the fibres of the wool, holding the mordant in place and causing the wool to accept the dye more readily. Different mordants enable different chemical reactions to take place between the dye and the mordanted wool, each resulting in a different colour.

Numerous colours can be obtained from one pot of dye by adding the mordant-treated yarn all together. If four mordanted skeins are used, plus one unmordanted skein, and all are simmered together, five different, usually related, colours will result. This is most useful if doing a project needing a number of colours, and very practical.

This is illustrated very clearly on page 33, in a picture of a pot of marigolds and skeins of wool showing the different colours that may be obtained at the same time.

Mordanting and dyeing must be done gently. *Simmering* is the key word. Simmering is a very slow gentle boil, any movement in the water being occasional and barely perceptible. Simmering slowly will cause the water to evaporate very little. Use a lid if you can to help retain the moisture. If you haven't a lid for the pot, a plate placed on the top will do the job of a lid.

Remember that wool will mat together if it is boiled violently, stirred briskly, or subjected to a sudden change of water temperature from hot to cold. Simmer, therefore, and *be gentle*.

Mordanting the fibre

Method	Ingredients		Time
MIX	specific mordant, e.g. alum, chrome, and cream of tartar if needed	in a little boiling water	till dissolved
MEASURE	water into pot and turn on heat		
ADD AND STIR	in the dissolved mordant		
ADD	wet wool to pot		
STIR GENTLY	bring to boil and simmer gently		for one hour or for time stated in mordant recipe
TAKE	wool out		and cool
RINSE	only if stated in recipe		
ROLL	in towel to absorb moisture		
DRY —— OR —— DYE	in shade immediately		

Mordant recipes

Follow these recipes carefully and don't use more mordant than specified.

Note: Use a plastic spoon and a small plastic container (e.g. a margarine tub) for mixing mordants, so that no other metal comes in contact with the mordant to cause a possibly adverse reaction. For the same reason I suggest plastic forks for moving the wool around in the pot while mordanting.

Alum

Alum is the most widely used mordant. It improves the brightness and fastness of the colours.

>   120 g wool tied into skeins
>   30 g alum (6 teaspoons)
>   7 g cream of tartar (two teaspoons)
>   4 litres water

Mix the alum and cream of tartar with a little boiling water. Stir and dissolve the chemicals well, then add to the rest of the water. Heat a little and then enter the wetted wool or fibres. Bring up to simmering point. Stir occasionally and continue simmering for 1 hour.

At the end of this time, take out the wool and cool. Don't wring or rinse. Just towel dry. Wool can then be dyed or dried and used later.

Note: An aluminium saucepan can act as a mordant, but the chemical gives much better results.

Chrome

Chrome brings out oranges and reds. The crystals are extremely sensitive to light.

>   120 g wool tied in skeins
>   3.5 g chrome (½ teaspoon)
>   4 litres water

For best results, mordant with chrome just before dyeing. Dissolve the chrome with a little boiling water, then add the rest of the water. Dissolve well and start to heat. Enter the wetted wool. Keep the fibres submerged in the pot. Always cover the pot with a lid or plate.

Bring slowly up to simmer and simmer for one hour, stirring gently from time to time.

Remove wool from pot and allow to cool. Do not wring or rinse wool, but wrap in a clean towel to absorb extra moisture. Use immediately for dyeing. If it can't be dyed immediately, dry the wool away from sunlight and store in a dark place for future use.

Copper

>   120 g wool tied in skeins
>   7 g copper sulphate (1½ teaspoons)
>   4 litres water

Dissolve copper in a little boiling water, then add the rest of the water. Stir well and enter the wetted wool. Bring slowly to the boil and simmer for one hour. Take out wool and cool. Don't wring or rinse, but wrap in a towel to absorb moisture. Dye immediately or dry away from sunlight and store in a dark place for future use.

Iron

Remember, too much iron hardens the wool.

>   120 g wool tied in skeins
>   3.5 g iron (ferrous sulphate) (½ teaspoon)
>   7 g cream of tartar (2 teaspoons)
>   4 litres water

Dissolve cream of tartar in a little boiling water, then add the iron. Add the rest of the water and stir well. Bring up to hand-heat temperature. Enter the wetted wool. Bring just to the boil and simmer for 15 minutes. Take wool out and cool. Rinse thoroughly in hot then cool water and wrap in towel to remove moisture. Use for dyeing immediately, or dry for later use.

Note: An iron saucepan can act as a mordant, but the chemical gives much better results.

Tin

>   120 g wool tied in skeins
>   3.5 g tin (stannous chloride) (1 teaspoon)
>   3.5 g cream of tartar (1 teaspoon)
>   4 litres water

Dissolve cream of tartar in boiling water, then add tin. Then add the rest of the water. Start to heat and add wetted wool. Simmer for one hour. Remove immediately and wash in warm soapy water. Then rinse. Wrap in a towel to remove moisture. Use for dyeing immediately or dry and use later.

Disposing of mordant residues

A recent finding by an Australian university was that the minute quantities of chemicals used in the process of home dyeing and mordanting were not seen to be dangerous to the environment because most of the chemical was absorbed into the wool. Any chemical residue left in the water was so minute that, when flushed away with other water, it created no problem at all. In other words, dilute any leftover liquid thoroughly and dispose of the waste water normally.

Eucalypt

The risks of mordanting

All the mordants mentioned in this book are poisonous, except for alum. They must be used with care. If proper precautions are taken, however, the risks are minimal.

ALWAYS

- Keep mordants away from children (pretty colours always attract children's interest).
- Avoid inhaling any powders.
- Use rubber gloves when handling mordants or mordant solutions.
- Wear a full apron.
- Use a mask if you are working with large quantities over a long period.
- Avoid eating or smoking while preparing or packaging mordants.
- Wipe over floors and benches after use with a wet cloth or mop.
- Remember, utensils used for mordanting must not be used for food preparation.
- Store mordants in a locked cupboard.
- Be sensible and don't take chances.

FIRST AID RULES FOR MORDANTS

- If inhaled, rest and keep warm. In severe cases, seek medical attention.
- In eyes: Rinse eyes continuously with water and seek medical attention.
- On skin: Wash with soap and water. In severe cases, seek medical attention.
- If swallowed, wash mouth thoroughly with water. Drink plenty of water. Seek medical attention. NOTE: With chrome, copper and iron mordants, induce vomiting (use Ipecac Syrup if available) and give 1% solution of bicarbonate of soda. Seek medical help IMMEDIATELY.

Come dye with me

Method for making dye and dyeing fibres

1. Collect dye stuff.
2. Note whether the plant material is fresh or dry; the part used (leaf, bark, etc,); the common name of the plant (add botanical name later); the place name and area of collection, and the season (autumn, winter, spring, summer).
3. Make yarn into skeins and tie loosely. If using fleece, tease out, that is, loosen fibres with your fingers. Handle carefully. Both fleece and yarn must be washed and rinsed clean.
4. Weigh the yarn on kitchen scales.
5. Weigh the dye material.
6. Note the time taken to obtain dye and the time taken to dye yarn.
7. Measure the water added to the dye pot.
8. Gently boil the dye stuff for the appropriate length of time (see page 15). When the colour looks deep enough, test with a little yarn. If the colour seems right, proceed.
9. Take out the plant matter from the dye pot. It's usually best to strain the liquid to remove any small bits which might cling to the wool. Cool the pot of dye liquid for 10 minutes before entering yarn or fleece, which has been wetted in warm to hot water.
10. Fleece must be handled gently at all times to prevent matting.
11. Simmer, and I stress, simmer, until the desired colour is obtained.
12. Rinse the wool in hot then cool water until water runs clear.
13. Dry in the shade.

Eucalyptus nicholii

Method for making dye and dyeing fibres

Method	Ingredients	Time
WEIGH	berries, leaves, flowers, etc.	Make notes as you proceed
MEASURE	water into pot	
BRING TO BOIL	the dye plant and water	Boil for allotted time
STRAIN TO REMOVE DYE PLANT	leaving only dye liquid in pot	Turn down heat
PUT IN WET WOOL	Stir gently	Simmer for allotted time or until desired colour is reached
TAKE OUT, RINSE AND DRY	in hot to cool water	

Dyeing with cochineal

Cochineal is the bright scarlet dye made from the crushed dried bodies of insects called *Dactylopius coccus* found on cactus plants grown in different countries in the world. The powdered form of the dye gives the best scarlet red, but the liquid can also be used. I used unmordanted wool and added a chemical aid in the form of stannous chloride (tin) to obtain the scarlet red colour. I found it necessary to use much more tin than I usually would. Very fine soft wool will take the dye more readily than coarse fibres.

Powder recipe

    75 g cochineal powder
    water
    5 g stannous chloride (tin)
    225 g wool or fibre

Mix cochineal powder with a little water. Boil gently for a few minutes and add enough water to cover the wool when it is added. Add the stannous chloride. Stir and dissolve. Now add the wool or other fibre (wetted in warm to hot water). The liquid in the dye pot should just cover the wool. Simmer for fifteen minutes or longer, just long enough to obtain the colour you want. Cool and rinse. Dry in shade.

You should get a scarlet colour using this method.

Liquid recipe

    25 ml cochineal (Queen Brand gives the best result)
    10 ml stannous chloride
    no water
    5 g wool

This basic recipe can be increased to accommodate the amount of wool you want to dye.

Place all together in pot and simmer ten minutes for a watermelon pink colour. Paler shades come from exhausting the dye pot.

Dyeing with lichen

Lichens are found growing on trees and fence posts. They are grey, sometimes with a tinge of light green, and peel off readily. There are quite a few kinds of lichen but to my knowledge only one produces the rich plum-coloured dye—the *Umbilicaria* lichen. The dye is contained in the orchil-producing acids found in the inner part of the lichen.

To test if a lichen has this dye, simply scrape off a small spot on the thin outer layer and place a drop of household bleach on the scraped surface. If an orchil-producing acid is present, the spot will turn pink, even red.

When collecting lichen, please remember that it takes many, many years to grow to decent-sized pieces, so don't strip a tree of all its lichen. Be selective; only take a little off each tree and only what is needed at that time. Lichen is a powerful dye—a handful is ample to dye one kilogram of wool or yarn.

To extract the dye, the lichen must be soaked in a solution of one part of ammonia to two parts of water. Place the handful of lichen in a large glass jar and cover with the water and ammonia solution. Leave it to stand for one to two weeks, or longer. It keeps for many months. Shake the mix from time to time. There is no need to remove the lid. If you have collected the right kind of lichen, you will see a reddish colour showing in the first few days of soaking.

When you are ready to dye, place the dye solution in a pot and start to heat. Remember ammonia has very strong fumes, even more so when it is heating. Don't go sniffing the pot. This kind of dyeing must be done in an open airy place, or with an electric fan operating. Add the yarn immediately. You will find that the yarn absorbs the dye quite quickly, so only leave as long as you need to get the colour you want. If you find that the dye is too strong a colour, add more water. Take out yarn and rinse.

Paler shades can be obtained by exhaust dyeing, placing yarn in a little at a time and absorbing the colour. Another way is to add more water to reduce the strength of the colour.

Lichen is a very strong dye in its original form. Shades from plum to pale lavender can be obtained.

If pinks are wanted, start with about one litre of stock dye, then add one or two tablespoons of white vinegar. Add more water, or use the exhaust method to lighten the colour. Shades from maroon to pale pink can thus be obtained.

No mordants are used with lichen dyes as the ammonia acts as a mordant. This colour holds quite well, although exposure to sunlight will fade it a little.

Come Dye in the Forest

Wood dyes: using timber shavings

Think of trees growing in the forest as coloured pencils waiting to be sharpened.

The colours extracted from timbers are incredibly beautiful when absorbed into fibres. Common trees such as Mulberry and Camphor Laurel will give muted golds, while exotic trees, such as Purple Heart from South America, will give rich greens, and Padauk from India, soft deep reds. To obtain timber shavings contact any woodcutter and he should readily supply you with numerous varieties.

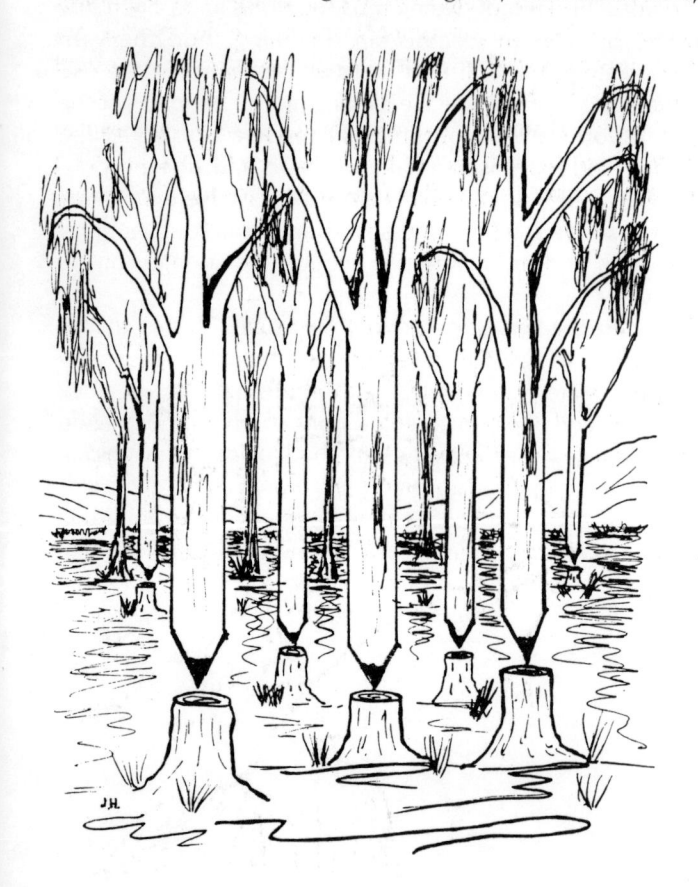

It is interesting to note that green timber shavings give a certain colour while the colour from seasoned timber will differ. Yet each sample of yarn dyed in this way will compliment each other. By using mordanted wools or yarn as instructed on page 24, and using green and seasoned shavings, you can obtain ten different colours from one variety of wood. Using the leaves, flowers, bark and nuts, a total list of 30 different colours can be obtained from one variety of tree, and all will compliment each other.

Pure silk or cotton yarn, fibre or material can also be used. See information on silk (pages 36 and 37) and on cotton (page 30).

The special index in the back of the book under 'Wood Shavings' will give you an indication of the colours you can obtain.

When using eucalypts you do not need to mordant to get a rich colour, but if you do you will obtain a good variety of shades.

I still firmly believe that mordanted yarns give the best results and that mordanting also helps to give a permanent colour. Always rinse the yarn well and shake out any residue of shavings that may adhere to the yarn.

Whether you live in Australia or overseas, each country has a variety of timbers with which to experiment. I found it exciting exploring different timber shavings, like Wenge from Africa which is a dark chocolate brown timber that gives beautiful chocolate brown colours on yarn. Purple Heart shavings, which were purplish in colour, gave amazing rich emerald greens on mordanted yarn. This was quite an unbelievable and exciting experience. Ebony was another interesting wood. It appeared to be both black and brown in a strip along the shaving. Most of the shavings I used were finely ground because the hardness of the wood cuts this way when using a lathe. It looked very much like fine tea leaves. But I never did try drinking the dye — it probably tasted dreadful! The yarn turned out lovely shades of rich chocolate brown. Brazilwood has been known for a long time as a dye source and it really is a rich source of soft coppery browns.

As I have said previously, each plant has a unique colour and it is so easy to obtain. It's the same with wood shavings, only it doesn't take as long to obtain the colours you want. Boiling the shavings for 15 minutes, then entering the mordanted yarn and simmering for 15 minutes gives you the rich colours as listed in the special index at the back of the book.

If you have any fragrant timber shavings, like Camphor Laurel or others that have not been used in the dyeing, put them in bowls in the house and allow the aroma to waft through on the breeze.

When you have finished using the wood shavings they can be recycled by using them in the garden or placing them around your pot plants. This stops the weeds from growing.

Tasmanian Blackwood

Queensland Maple

Cotton on to Dyeing

This process can be used for cotton yarn, in skeins or cotton material. Cotton is a vegetable fibre. It can be used in plant dyeing but the results are paler.

To get the best results, it's better if you mordant the cotton with alum first. For example, if you are using 120 g of cotton material or yarn you will need:

    4.5 litres water
    7 g washing soda (2 teaspoons)
    30 g alum (1.5 tablespoons)

First of all, wash the cotton material or yarn thoroughly to remove any sizing or other treatment that may be in the fabric. Rinse well then continue with mordanting.

Dissolve the washing soda and alum in a pot with a little boiling water, then add the rest of the water and stir. Add the clean wet cotton material and bring to boil slowly, giving it an occasional stir.

Boil for one hour and leave to soak in the pot overnight. Then squeeze out excess liquid and go ahead with the dyeing process.

Red Cedar

Ironbark

Above: Willow peppermint gum (*Eucalyptus nicholii*) These fine blue-green leaves (note the orange stems) are a valuable dye source. This eucalypt grows to about 25 m. The dye from it can colour fibres ginger to brownish black, depending on the mordant used

Right above: Handspun plant-dyed shawl collar handknitted by Karlene Lewis. Dyes are from *Eucalyptus nicholii*

Right: *Jacaranda mimosaefolia* The beautiful jacaranda, which is covered in fat mauve-blue tubular flowers in late spring, is useful to the dyer, the flowers providing colours ranging from fawnish grey to olive green

Below: Flame tree (*Brachychiton acerifolium*) Aptly named, as the tree appears as a red flame which can be seen for quite a distance amongst the greens of other trees. All the leaves drop as the tree comes into flower. The flowers themselves are in clusters, small and waxy and bell-shaped. Not a spectacular dye source, but it does give a pinky-fawn colour to unmordanted yarn

Right: Lemon-scented tea tree *(Leptospermum petersonii)* This tree's leaves give off a delightful lemon fragrance when crushed. It grows to about 4 m high and has white flowers. The dye from the leaves will dye fibres fawn, golden olive, greenish gold and soft brown

Below: *Lichen umbilicaria* Lichens grow on fences, trees and rocks. This particular lichen, usually found on trees, resembles greyish curled leaves. Soaked in ammonia and water for a week or two the lichen releases a strong purple/red dye, which will dye fibres shades ranging from lavender to plum. A pink shade can also be obtained

Below left: Handspun, hand-crocheted teacosy by Karlene Lewis, the pink colour obtained from lichen

Below right: Lichen-dyed handspun, handwoven shawl by Karlene Lewis, modelled by Sawako Kawakami

Left: French marigolds *(Tagetes patula)* Pom-pom-like annual flowers which come in shades of yellow, gold and bronze. They are excellent for dyeing, resulting in a number of quite distinctive colours with different mordants. The range is gold, ginger, golden brown and deep olive

Wool yarn being dyed with flowers of the French marigold. Each piece of wool, mordanted with a different mordant, is taking on a different shade. Clockwise from the top, the mordants are alum, none, copper, chrome, iron (in the centre)

Above: Brown onion skins *(Allium cepa)* The papery golden brown skins of the onion are useful to the dyer. They easily impart a golden dye in hot water, which will dye yarn shades of ginger and yellow

Right: Handspun wool vest dyed with onion skins is by Karlene Lewis. Handbag is plant-dyed wool needlepoint embroidery, also by Karlene Lewis

33

Above: Pawpaw *(Carica papaya)* This tree is found in most household gardens in tropical and sub-tropical areas. Use the leaves as a dye source to dye yarn yellow, green, grey and cream

Left above: Passionflower *(Passiflora sp.)* This inedible member of the passionfruit family is very fast growing. The leaves yield a dye that will colour yarn yellow, green and olive

Left: Plumbago or leadwort *(Plumbago auriculata)* The sky blue or white flowers of this shrub stand out well against its mid-green foliage. Gold, brown and grey colours can be obtained from this plant

Below Most people are familiar with roses. The petals are used in potpourri, but to the dyer they are useful also. Dye from them will colour yarns lime green, olive green and soft brown. The leaves give similar colours

Method for making dye from wood shavings and dyeing wool

Method	Ingredients	Time
WEIGH	Wood shavings	Make notes as you proceed
PLACE	Wood shavings in jar	
MEASURE	Water and pour into jar with shavings	Allow to stand for *24 hours or longer*
STRAIN LIQUID THROUGH NYLON NET INTO POT	Tie up shavings in net. Place in pot, bring to boil	Boil for *15 mins*
TURN DOWN HEAT AND PLACE WET WOOL IN POT	Stir gently	Simmer for *10-15 mins*
REMOVE WOOL, RINSE AND DRY	in hot to cool water	Dry in the shade

Dyeing for Silk

Silk is an exquisite fibre. Its rich lustre and soft texture create images of sheer opulence.

It is formed by the silkworm, the caterpillar of the silk moth, which feeds mainly on mulberry leaves. When the caterpillar is fully grown, it expels a fine, soft filament (silk) and spins a cocoon around itself in a continuous series of figure eights. The fibre is formed from a liquid protein called fibroin which is secreted in separate streams by two glands in the silkworm's body. Two adjacent glands secrete a substance called sericin which cements the two protein secretions together as they are expelled through spinnerets in the silkworm's mouth. The protein solidifies to form a triangular-shaped, smooth, translucent fibre. The sericin also acts as a cement to hold the fibre in place on the cocoon. The fibre around the cocoon is a single strand which measures up to about 1200 metres in length.

In order to unwind the silk filament from the cocoons, they are placed in boiling water to soften the sericin. The filament is then pulled from the cocoon and wound onto a reel. The filaments are spun into yarn and then skeined. The skeins are degummed by boiling to remove the last traces of sericin. They are then ready for dyeing.

If you buy pure silk fibre to spin, or material, you will find it is ready for dyeing and the only preparation that is needed is a light washing, using a product that is suitable for washing wool. Handle with care and don't use water that is too hot.

It is easier to dye if wound into skeins, although it can be dyed loose if handled with care. Pure silk material can also be dyed. It readily accepts the dyes from plants and wood shavings.

If using eucalypts, no mordanting is necessary but if other plants or wood shavings are used, mordanting can be done to achieve other colours. Usually they are paler than for wool. Follow the mordanting process as for wool at the front of this book (pages 19-20).

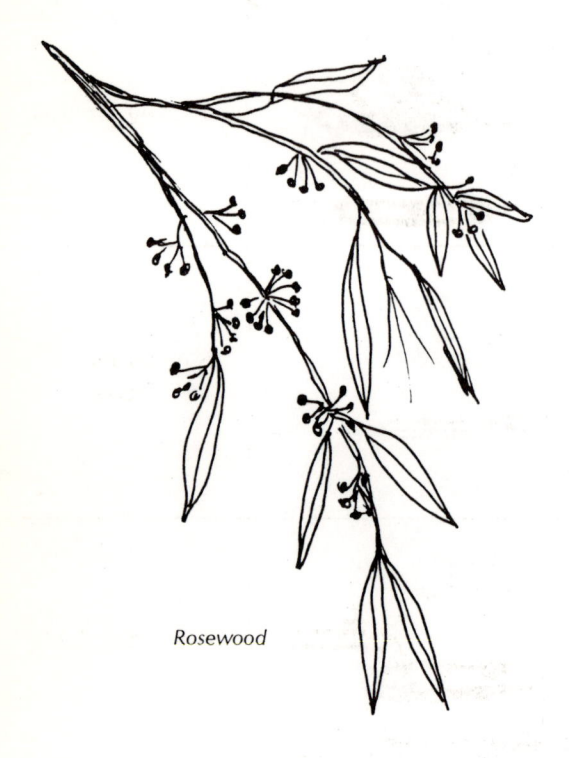

Rosewood

Black Bean

Method for making dye from wood shavings and dyeing silk

Method	Ingredients	Time
WEIGH	Wood shavings	Make notes as you proceed
PLACE	Wood shavings in jar	
MEASURE	Water and pour into jar with shavings	Allow to stand for *24 hours or longer*
STRAIN LIQUID THROUGH NYLON NET INTO POT	Tie up shavings in net. Place in pot, bring to boil	Boil for *15 mins*
TURN DOWN HEAT AND PLACE WET SILK IN POT	Stir gently	Simmer for *10-15 mins*
REMOVE SILK, RINSE AND DRY	in hot to cool water	Dry in the shade

Dyeing for children

Children can have a lot of fun dyeing, as long as it's safe. *No* mordants or chemicals are used, but supervision is needed wherever children and pots of hot liquid are concerned. Who knows—an interest developed in the craft while they are young may carry through to a lifelong involvement.

These materials give good results unmordanted:

Brown onion skins—gold to ginger colours
Mulberries—purpley grey
Nasturtium flowers—soft green
Bracken fern—pale green
Eucalyptus leaves (*E. cinerea, E. nicholii*, or any other, for that matter)—colours range from cream to yellow, apricot to rust

Check the colour index under the 'unmordanted' heading for other plants they might like to try—but don't let them use poisonous plants.

For young children take a shortcut to speed the process and retain their interest. Put the dye matter into a pot, cover with water, add the wetted white wool and simmer all together. Check after twenty minutes for the depth of colour. Eucalyptus leaves take a little longer than the others mentioned to develop their colour.

Don't overcook. Twenty minutes is enough time to develop the colour needed from flowers, and from some leaves and berries also. Eucalyptus leaves can be simmered for up to two hours if they have the patience to wait that long. The longer you leave the eucalyptus leaves, the deeper the colours will be. Then just rinse the yarn in the usual way and dry in a shady spot.

They can use their dyed wool for all manner of projects—knitted dolls' clothes, crochet, weavings knitted squares for pincushion gifts.

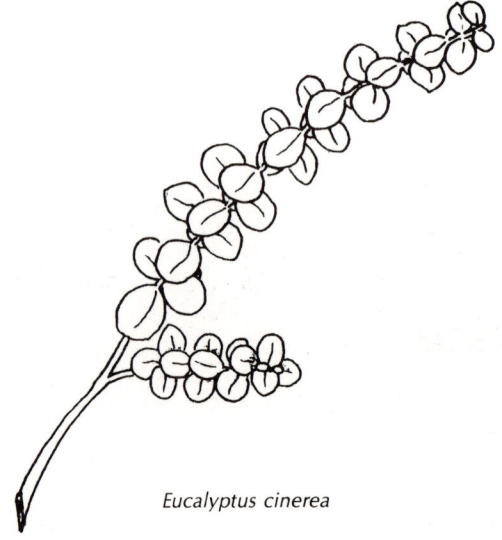

Eucalyptus cinerea

Are you dyeing to know?

Here are a few useful hints:
- Open fires are not good as a source of heat because you can't regulate them and keep an even heat.
- Colours can be mixed. Always try a little together first before pouring one pot of dye into another, even putting in a tiny sample of wool to get an idea of the colour.
- Plants can be used dried or fresh. There may be a slight difference but the same tone is usually obtained.
- Berries which may only be available in small quantities can be saved and stored (properly marked, especially if poisonous) in a freezer for future use. When you have enough, just defrost and start your dyeing.
- Many herbs can be used as dyes.
- When collecting lichen, take a small sharp knife with you and a small bottle of bleach. Collect a sample of lichen and scratch the surface. Put a drop of bleach on it, mixing a little for a few moments. This should show as pale pink if you have the right lichen. You can judge from that if it is worth collecting. Always collect lichen very carefully, as most take many years to grow. Always collect selectively, taking a little from this tree and a little from that tree. This is always good advice when collecting leaves and flowers, too. Always prune a tree in the proper way.
- Even though it is good to be accurate with the dyeing process, don't become fanatical about a stalk or two more or less. It is not a life or death situation, you're the only one dyeing! It's a fabulous craft and it should be fun.
- Test skeins for a new plant need only be a few metres long. Eight times around hand to elbow makes a good-sized skein.
- Here is an interesting exercise.
Mordant 2 metres of thick white wool yarn in alum, another 2 metres in chrome, another 2 metres in copper, and a further 2 metres in iron. Dry the yarn after the mordanting process. Then join the four lengths of yarn into one continuous length. Using a pair of large knitting needles (approximately 6 mm), cast on twenty stitches. Continue knitting until you have used up all the yarn to make a small square. Wet the piece and then pop it into a dye pot. Hey presto! A striped square with beautifully blended colours which could be used as a potholder. Using more yarn, you could make a cushion cover.

Westringia fruticosa

Making a dye sample book

Small pieces of yarn dyed all sorts of colours left in a bag aren't very attractive, and end up being very confusing. Do yourself a favour and make a sample book.

You will need a large two or three ring binder, some white cardboard, scissors, ruler, pencil and a punch to make the holes for the yarn. First, cut the cardboard to fit the binder, but instead of cutting each piece the same width, stagger the sizes.

Let's say the binder is 35 × 25 cm: cut the first piece of cardboard 35 × 8 cm; the second piece 35 × 10 cm; the third piece 35 × 12 cm; the fourth piece 35 × 14 cm.

Then punch two or three holes in the left side to fit the binder.

You then need to punch a series of holes, in groups of five, down the other side of the cardboard 1 cm in from the edge. With a binder 35 cm deep, you could punch 25 holes down the side of each piece of cardboard, marking each in turn A (for alum), CH (for chrome), C (for copper), I (for iron) and UM (for unmordanted). You will then be able to house 25 samples from five different plants on each page. A small piece of dyed yarn, say 16 cm in length, is folded in two and looped through the hole in the cardboard.

The advantage of staggering the width of the pages is that on four pages you will have displayed at a glance a full range of colours from twenty different plants, shown to perfection, overlapping each other only a fraction. The knotted parts of the wool samples don't overlap so lots of pages can be put in the binder. I have sixteen pages made this way in one of my books and it isn't fat and cumbersome. If you want to show more details, add a pressed sample of each plant to the book. A photo of the plant also helps in identification. All you need to do now is write in each plant name next to the samples. A photograph of a dye sample book appears inside the back cover.

Pressing plants

I always select a perfect specimen piece from each plant I collect and press it, handy for future reference or for your sample book. It need only be a small piece, e.g. a leaf or bunch of leaves, depending on size, a flower or piece of bark. Berries are not the best to press (very messy), but it can be done.

Pick flowers or leaves in peak condition, never after rain, as the moisture in them will spoil them as they dry and they may become mildewed on the petals or leaves. Collect on a dry sunny day and nip off the stalks of the flower heads if they are thick. Arrange the flowers in an open way on some blotting paper with newspaper underneath. Smooth leaves out flat. Cover with another layer of blotting paper and newspaper. Place the 'sandwich' between two heavy books and weigh down with two bricks. If you are lucky enough to have a flower press, all you need is some blotting paper and newspaper.

With berries, either squash them whole or slice them in half and place between blotting paper and tissues in a flower press or between two heavy books.

Don't use paper towel instead of blotting paper as it is usually self-patterned and can leave marks on petals or soft leaves. Put a slip of paper in next to the specimen with the name clearly marked. Leave the specimen plant under pressure for at least a month in a light, sunny, airy room.

Yellow and orange coloured plants or flowers keep their colours well when pressed. Even though rose petals may turn cream, don't be put off trying. Grey leaves remain grey.

It will help if, after you have pressed the specimen, you place it on a suitable background. Pale coloured plants look great on a dark background and vice versa. Then completely seal over the specimen with clear self-adhesive vinyl. This will make a neat card specimen for your sample book or to file in a box for future reference.

Method for making a dye sample book

Method	Items
2 or 3 ring binder, sheet of white cardboard, punch, scissors, ruler, pencil	yarn
Measure cardboard, cutting each piece 2 cm wider than previous piece	
Punch holes for binder on left of card; on right edge punch 25 holes 1 cm in from edge and 1 cm apart for yarn samples	
Mark page with letters as shown, plant name, collection place and date	plant name → A for alum / CH for chrome / C for copper / I for iron / UM for unmordanted — place and date
Cut dyed yarn 16 cm long; fold each piece and loop through hole	16 cm
Place completed pages in binder so that each page overlaps the next	

Uses for dye samples

If you frequently dye only small lots of yarn, you will find after a while you are left with bags of little coloured bits. Here are some uses for dye samples and left-over bits.
- Make an attractive cushion cover by joining blending colours and crocheting them into a large square. Make two squares and presto, you have a cushion cover.
- Make a rug for the car by knitting or crocheting the pieces into manageable squares and joining them together.
- Make a small wall hanging—knit or weave blending colours into a rectangle and attach it to dowelling rods.
- A larger wall hanging can be made by crocheting blending colours together into the size you want. Dowelling rods can then be attached for hanging.
- Make up an interesting design on paper, then weave a copy in a miniature square, say 20 cm × 20 cm. Make several, which can be mounted on stiff board and hung in groups of two, three or more.
- Knit or crochet two small squares (8 cm × 8 cm), sew together and stuff with raw fleece for a pincushion. (The raw fleece keeps the pins from rusting.)

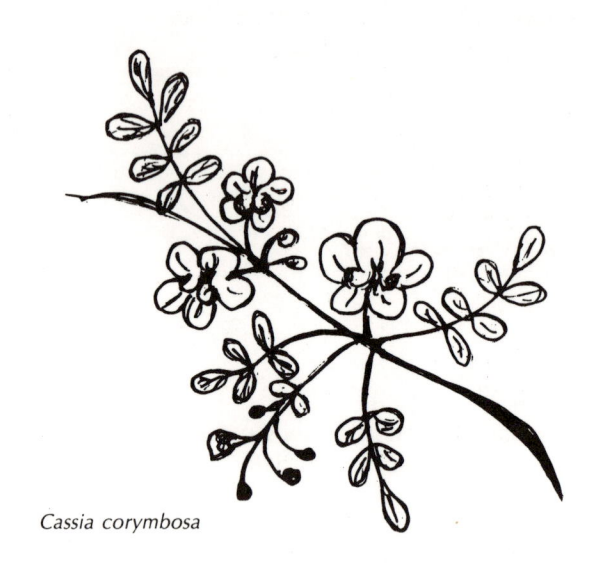

Cassia corymbosa

Storing dyestuffs

Dried leaves, flowers and bark can be stored in paper bags or cardboard boxes. Only use plastic bags when you are sure the dye stuff is completely dry. Always mark bags with the type of plant they contain as after they are dried they can be quite difficult to identify. Large glass jars are also useful for storage, and will look neater if you are using part of your pantry. Leaves can be stored indefinitely if dried out completely, and dried flowers can also be kept for quite long periods.

Berries can be stored in the freezer for up to six months in plastic icecream containers. They will deteriorate after that. Mark the containers clearly, putting on the date of collection and an expiry date.

I'm dyeing to tell you about...

A friend and I were doing some dyeing in her kitchen one rainy day. We were working with pine needles, cassia plant and eucalyptus leaves. Her cat took an interest in the pots of dye cooling on the back stairs, but after sniffing the first pot, a puzzled look came over his face. After sniffing the third pot, he stood up, shook himself all over and *raced* through the house and into the rain in the front yard, where he stayed until we had finished. Poor puss!

Her husband came home soon after and looked in our pots on the stove. He hoped it wasn't *his* tea we were cooking. I must admit some strange odours were coming out of the kitchen.

I decided it was time I went home.

★

Another time, my Labrador dog gave me a wide berth when I started dyeing in the courtyard. He usually sat wherever I was working but he must have reckoned I could dye alone that day as he took off and parked himself elsewhere.

★

My son Paul always called me the old witch when he came home from school and found me dyeing. He actually made me a witch's hat to wear on one occasion.

★

Ever have someone give you flowers which after a week you've had to toss out as they'd come to their end? In the future, don't toss them out—let the petals dry and store them for your next dyeing experiment. I did just that once. I had been given a bunch of red roses and a bunch of pink roses. With these petals, and wool mordanted in copper, iron, chrome and alum, I got a lovely range of colours from soft brown to a number of lovely greens.

★

In another dye pot I had *Eucalyptus cinerea* and with the same mordants came rich rust reds to a deep copper colour. I also had some stored marigold petals, which when used for dye resulted in pale golds. The problem with these flowers was the smell! I believe marigolds are sometimes known as 'Stinking Roger'—well, they live up to that name! Fortunately I was dyeing on a gas stove out in the courtyard,. You would have been forgiven for thinking that there was a problem with the septic system!

★

I experimented with red berries from a plant in my front garden. It was *Solanum seaforthianum* (potato vine). The wool samples turned out a nice apricot colour with alum-mordanted wool.

I tried dyeing with parsley once, but the results were so disappointing I put the samples in with some *Eucalyptus cinerea* and obtained a nice range of rust shades. So don't despair if something gives you a disappointing colour—just put it in another dye pot.

I tried parsley again another time, using a copper mordant, and got a nice green.

★

On another dyeing day, I got an early start at 9.30am with the family out of the way. Firstly, brought upstairs enough rain water to do all the dyeing. Dyestuffs ready: jacaranda flowers, elkhorn leaves, pomegranate flowers, passionfruit skins, Buderim soil (red).

Put pomegranate flowers in an enamel pot with rain water and—good grief! forgot to weigh the dried flowers! Oh well, too late! Will check on that later when I collect some more. I had these growing in the front garden. As the water started to heat, the dye released was very thick (inky brown). Putrid smell! Kitchen fan on full blast to disperse the smell. Mordants used were alum, oxalic acid, chrome, tin and copper.

First distraction was a friend arriving with a mohair fleece. Checked on pots then tipped the fleece out on the patio. Sawdust everywhere. (The goats had been housed in a shed with a sawdust floor.) Too bad. Sweep it into the garden. One has to be eccentric to take on dyeing at home. Good friend didn't say anything about the smell coming from the kitchen. Nice to know that friends accept me as I am. Quite mad at times, especially when I'm dyeing.

Elkhorn leaves were next on the list for dyeing. I had been soaking the leaves in ammonia and water for some months to release the dye. This couldn't be weighed because of the liquid. Added a small amount of water and started heating the lot. Yuk! What a pong! (Heavy ammonia smell.)

Knock! Knock! Someone at the door. 'Yes, Harvey, what can I do for you?' Before he answers, the question was, 'What's burning?' 'Nothing Harv, I'm only dyeing.' Puzzled look—ask a silly question, get a silly answer. He collected what he wanted and was off. Peace at last. Back to the dyeing. Mordants used alum, chrome, copper. No mordant in one sample.

Jacaranda flowers had been collected freshly fallen in November and allowed to dry out. I remembered to weigh the flowers this time. Placed them in an enamel pot with rain water and started to simmer. This didn't look very exciting—not much colour. Mordants used were oxalic acid, cream of tartar and tin, alum, chrome and copper.

I must mention that my dog Julius Caesar, who loves to be with me no matter what I am doing, couldn't stand the smell upstairs any longer. He made a fast exit downstairs and stayed there until I finished dyeing. Poor chap.

Passionfruit skins which had been kept in the freezer for about six months. Started to simmer, not a lot of colour but it may give interesting results. Mordants were cream of tartar and tin, oxalic acid, chrome and copper.

Buderim soil was the next on the list. Maybe a pink from this? This soil stains one's socks, so why not wool? Mordants used were alum, copper, chrome and no mordant on one. I had a six-litre icecream container three-quarters filled with soil so added water and strained off liquid. Started simmering with a handful of soil tied in a rag plus liquid in pot.

The results of these experiments were:

Pomegranate flowers: All samples came out gold-ginger colour but each a shade different owing to the different mordants used.

Elkhorn leaves: All colours were a little disappointing, most being in the beige-grey colour range.

Jacaranda flowers: One came out a nice leaf green and the others slightly greenish.

Passionfruit skins: One was a nice grey-mauve, another green-grey and the rest greyish.

Buderim soil: A nice pale peach colour with the alum and unmordanted wool and the others beige.

All dyestuffs were simmered one hour before the wool was added, then simmered for another half hour. Even though the results weren't spectacular, I enjoyed my dyeing day and will experiment again another time with other dyestuffs.

Lists

1. COLOUR LIST

Cream, Lemon, Yellow, Gold
Apricot, Rust, Ginger, Orange, String
Fawn, Brown
Grey, Black
Green, Olive
Scarlet, Red, Pink, Watermelon, Plum, Purple, Lavender

2. PLANTS FOR COLOURS LIST

3. BOTANICAL TO COMMON NAMES LIST

4. WOOD SHAVINGS FOR COLOURS LIST

HOW TO USE THE LISTS

This cross reference is very simple.

1. Firstly, choose the colour you want to obtain within this range of shades: Cream, Lemon, Yellow, Gold, String, Apricot, Rust, Ginger, Orange, Fawn, Brown, Grey, Black, Green, Olive, Scarlet, Red, Pink, Watermelon, Plum, Purple and Lavender.

Note: Blues are not readily found in Australian plants. Some plants give blueish greys or lavender blues, but a true blue is yet to be discovered.

2. Turn the page for the start of the Colour List.

This will give you the common name of the plant you need to obtain that colour. You may have some of these plants in your garden or know where you can get them.

3. Next turn to the Plants for Colours List. This is in alphabetical order and following this through will give you

(a) botanical name
(b) plant part used
(c) whether alum (potassium aluminium sulphate) is used
(d) whether chrome (potassium dichromate) is used
(e) whether copper (copper sulphate) is used
(f) whether iron (ferrous sulphate) is used
(g) whether no mordant is needed
(h) whether tin (stannous chloride) is added.

You now should have the information you need to start the dyeing process. I have also added a Botanical to Common Name List so that those of you who are familiar with botanical names will then be able to find the common name under which the plant is listed.

1 Colour List

CREAM

Common name	Plant part
Agapanthus	Flowers
Azalea	Pink flowers
Banksia, Hill	Leaves
Basil	Leaves
Camphor laurel	Wood shavings
Chrysanthemum	Leaves
Clerodendron	Flowers
Comfrey	Leaves
Daisy tree, Mexican	Leaves
Elkhorn	Leaves
Fennel	Flowers
Flame tree	Flowers
French marigold	Flowers
Fungi	Whole plant
Grape	Leaves
Grey box	Leaves
Gum, Lemon-scented	Leaves
Hackberry, Chinese	Leaves
Hibiscus	Leaves
Hibiscus, Orange	Flowers
Hill banksia	Leaves
Jasmine	Flowers & leaves
Lasiandra	Leaves & flowers
Lemon-scented gum	Leaves
Lichen, Old man's beard	Whole
Marigold, French	Flowers
Mexican daisy tree	Leaves
Mistletoe	Leaves
Moses-in-a-boat/basket	Leaves
Nasturtium	Flowers
Old man's beard	Whole
Passionfruit	Skins
Pawpaw	Leaves
Polygala	Flowers
Potatoes, red sweet	Root
Purple King beans	Roots, stems, seeds
Sandalwood	Leaves
Spider plant	Leaves
Wattle, White sallow	Leaves
Westringia	Flowers & leaves
White cedar	Berries

YELLOW

Common name	Plant part
Azalea	Pink flowers
Banksia, Hill	Leaves
Basil	Leaves
Bleeding heart	Flowers
Blue gum	Leaves
Chrysanthemum	Flowers & leaves
Coral tree	Leaves
Dahlia	Flowers
Dogwood	Leaves
Fennel	Flowers
Gum, Lemon-scented	Leaves
Gum, blue	Leaves
Hill banksia	Leaves
Jasmine	Flowers & leaves
Lasiandra	Leaves & flowers
Lemon-scented gum	Leaves
Onions, Brown	Skins
Passion flower	Leaves
Pawpaw	Leaves
Privet	Leaves
Purple King beans	Stems
Rosemary, Coastal	Leaves & flowers
Turmeric	Roots
Wattle, White sallow	Leaves

LEMON

Common name	Plant part
Banksia, Hill	Leaves
Daisy tree, Mexican	Leaves
Hill banksia	Leaves
Mexican daisy tree	Leaves

GOLD

Common name	Plant part
Aluminium plant	Leaves
Beetroot	Root
Bixa	Seeds
Bottlebrush	Leaves
Camphor laurel	Wood shavings & leaves
Cassia	Flowers
Chrysanthemum	Flowers & leaves
Coffee tree	Leaves
Coral vine	Flowers
Everlasting daisy	Flowers

Left: Silky oak *(Grevillea robusta)* This tree, one of the largest of the Grevillea family, grows to about 30 m. Its timber is used in cabinet making. Its golden orange flowers appear in late spring and early summer. Yarn can be dyed a greenish gold from the leaves

Below: Tree ferns *(Cyathea australis)* The beautiful soft green fronds are fresh and cool looking. Tree ferns grow to great heights in the rainforests but in the home garden they rarely grow beyond 3 m. Pretty shades of green can be obtained from the dye of these plants

Above: *Vitex purpurea trifolia* These unusual shrubs, whose leaves are grey one side and purple the other, can get quite rambly if left unpruned. A dye from the leaves can colour yarn soft grey-green

Right: *Westringia fruticosa* Common all along the east coast of Australia, this shrub has deep green tiny leaves and mauve or white flowers about the size of a 5 cent piece. This plant gives shades of pale yellow, olive green, grey-green and greenish cream

Detail of jacket by Judith Hallett. The leaf design is knitted in beetroot-dyed wool

Detail of wool bargello wall picture by Pat Burgess. Dyes from *Acacia saligna* wattle blossom. Green with copper mordant; creams and gold with tin mordant; black and dark grey with iron mordant; brown with chrome mordant. Background with copper and chrome mordant

Detail of wool scarf by Pat Burgess, using dyes from wattle blossom, mainly *Acacia macradenia* (zig-zag wattle). The labels on the horizontal stripes read (top to bottom) chrome, tin, copper. Other mordants were also used, and the stripe below the tin was unmordanted

Granny square rug in plant-dyed wools

48

Common name	Plant part
Flame tree	Flowers
French marigold	Flowers
Kalanoa	Leaves
Lemon-scented teatree	Leaves
Lichen, Old man's beard	Whole
Marigold, French	Flowers
Old man's beard	Whole
Plumbago	Flowers & leaves
Pomegranate	Flowers
Prickly pear	Fruit
Rose	Petals
Scaley buttons	Flowers & leaves
Silky oak	Leaves
Tallowwood	Wood chips & leaves
Teatree, Lemon-scented	Leaves
Verbena, Veined	Flowers
Verbena	Leaves
Wattle, Brisbane	Leaves
Wattle, Zig-zag, Flat, Clay	Leaves
Westringia	Leaves

FAWN

Common name	Plant part
Agapanthus	Flowers
Allamanda	Flowers
Aluminium plant	Leaves
Banksia, Hill	Flowers
Bloodwood, Swamp, Gum	Leaves
Brazilian cherry	Berries
Buderim soil	Soil
Camellia	Dark red flowers
Carnations	Flowers
Coffee tree	Leaves & berries
Coral tree	Flowers
Cotoneaster	Leaves
Cypress, Monterey	Leaves
Dogwood	Leaves
Dracaena	Leaves
Flame tree	Flowers
Gum, Lemon-scented	Leaves
Gum, Narrow-leaf red ironbark	Leaves
Hibiscus	Leaves
Hill banksia	Flowers
Ironbark, narrow-leaf red	Leaves
Jacaranda	Flowers
Lemon-scented gum	Leaves
Lemon-scented teatree	Leaves
Lichen, Old man's beard	Whole
Loquat	Leaves
Mistletoe	Leaves
Narrow-leaf red ironbark	Leaves
Oak	Acorns
Old man's beard	Whole
Pine needles	Needles
Plumbago	Flowers & leaves
Potato vine	Berries
Purple King beans	Seeds
Red salvia	Flowers
Scarlet sage	Flowers
Spider plant	Leaves
Swamp bloodwood gum	Leaves
Sweet pea	Flowers & leaves
Sword fern	Leaves
Teatree, Lemon-scented	Leaves
Wattle, Brisbane	Leaves
Wattle, Silver	Leaves
Wattle, Zig-zag, Flat, Clay	Leaves

BROWN

Common name	Plant part
Aluminium plant	Leaves
Banksia, Hill	Leaves
Bixa	Seeds
Bottlebrush	Leaves
Brazilian cherry	Berries
Camphor laurel	Leaves
Canna	Leaves
Cobbler's pegs	Seeds
Coffee tree	Leaves & berries
Coral vine	Flowers
Everlasting daisy	Flowers
French marigold	Flowers
Geranium, Red	Flowers
Gum, Lemon-scented	Leaves
Hill banksia	Leaves
Lemon-scented gum	Leaves
Lemon-scented teatree	Leaves
Lichen, Old man's beard	Whole
Loquat	Leaves
Marigold, French	Flowers
Mistletoe	Leaves
Oak	Acorns
Old man's beard	Whole
Pine needles	Needles
Plumbago	Flowers & leaves
Pomegranate	Flowers
Purple cabbage	Leaves
Rose	Petals & leaves
Tallowwood	Wood chips & leaves
Teatree, Lemon-scented	Leaves
Wattle, Brisbane	Leaves
Wattle, Silver	Leaves
Wattle, Zig-zag, Flat, Clay	Leaves

GREY

Common name	Plant part
Aluminium plant	Leaves
Ardisia	Berries
Azalea	Pink flowers
Banksia, Hill	Flowers
Basil	Leaves
Blackberry nightshade	Berries
Blue gum	Leaves
Blueberries	Fruit
Bottlebrush	Leaves
Camellia	Dark red flowers
Camphor laurel	Wood shavings
Clerodendron	Flowers
Comfrey	Leaves
Coral tree	Leaves & flowers
Deadly nightshade	Berries
Dogwood	Leaves
Elkhorn	Leaves
Everlasting daisy	Flowers
Flame tree	Flowers
Grey box	Leaves
Gum, Blue	Leaves
Gum, Narrow-leaf red ironbark	Leaves
Hibiscus	Leaves
Hill banksia	Flowers
Ironbark, Narrow-leaf red	Leaves
Jasmine	Flowers
Lasiandra	Flowers
Moses-in-a-boat/basket	Leaves
Narrow-leaf red ironbark	Leaves
Passionfruit	Skins
Pawpaw	Leaves
Pine tree	Cones
Plumbago	Flowers & leaves
Potatoes, Red sweet	Root
Purple King beans	Seeds
Purple cabbage	Leaves
Red salvia	Flowers
Scaley buttons	Leaves
Scarlet sage	Flowers
Spider plant	Leaves
Verbena, Veined	Flowers
Verbena	Leaves
Wattle, White sallow	Leaves
Westringia	Flowers & leaves

BLACK

Common name	Plant part
Coffee tree	Leaves
Gum, Willow peppermint	Leaves
Peppermint gum, Black	Leaves
Rosemary, Coastal	Leaves & flowers
Scaley buttons	Flowers
Willow peppermint gum	Leaves

GREEN

Common name	Plant part
Asparagus fern	Leaves
Azalea	Pink flowers
Banksia, Hill	Leaves
Blue gum	Leaves
Bottlebrush	Leaves
Bougainvillea	Flowers & leaves
Bracken fern	Leaves
Camphor laurel	Wood shavings
Cassia	Leaves
Celery	Leaves
Chrysanthemum	Flowers & leaves
Clerodendron	Flowers
Cobbler's pegs	Seeds
Coral tree	Leaves
Daisy tree, Mexican	Leaves
Elkhorn	Leaves
Fennel	Flowers & leaves
Fishbone fern	Leaves
Geranium, Red	Flowers
Grass	Leaves
Grey box	Leaves
Gum, Blue	Leaves
Gum, Lemon-scented	Leaves
Gum, Narrow-leaf red ironbark	Leaves
Hackberry, Chinese	Leaves
Hibiscus, Orange	Flowers
Hill banksia	Leaves
Ironbark, Narrow-leaf red	Leaves
Jacaranda	Flowers
Jasmine	Flowers & leaves
Lasiandra	Leaves
Lemon-scented gum	Leaves
Lettuce	Leaves
Mango	Leaves
Mexican daisy tree	Leaves
Mint	Leaves
Moses-in-a-boat/basket	Leaves
Narrow-leaf red ironbark	Leaves
Nasturtium	Flowers & leaves
Parsley	Leaves
Passionfruit	Skins
Passion flower	Leaves
Pawpaw	Leaves
Privet	Leaves
Purple King bean	Stems
Rosemary, Coastal	Leaves & flowers
Rose	Petals & leaves

Common name	Plant part
Sandalwood	Leaves
Scaley buttons	Leaves
Spider plant	Leaves
Sweet pea	Flowers & leaves
Sword fern	Leaves
Tallowwood	Wood chips
Tree fern, Rough	Leaves
Turmeric	Root
Verbena, Veined	Flowers
Verbena	Leaves
Vitex	Leaves
Wandering Jew	Leaves
Westringia	Flowers & leaves
Yesterday, today, tomorrow plant	Flowers

OLIVE

Common name	Plant part
Agapanthus	Flowers
Banksia, Hill	Flowers
Basil	Leaves
Bleeding heart	Flowers
Blue gum	Leaves
French marigold	Flowers
Gum, Blue	Leaves
Hibiscus	Leaves
Hill banksia	Flowers
Lasiandra	Flowers & leaves
Lemon-scented teatree	Leaves
Marigold, French	Flowers
Oak	Acorns
Purple King beans	Seeds
Teatree, Lemon-scented	Leaves
Westringia	Flowers & leaves
Yesterday, today, tomorrow plant	Flowers

APRICOT

Common name	Plant part
Buderim soil	Soil
Cobbler's pegs	Seeds
Peppermint gum, Black	Leaves
Potato vine	Berries
Tallowwood	Leaves

RUST

Common name	Plant part
Argyle apple	Leaves
Gum, Argyle apple	Leaves
Gum, Mealy stringybark	Leaves
Mealy stringybark	Leaves
Peppermint gum, Black	Leaves

GINGER

Common name	Plant part
French marigold	Flowers
Gum, Willow peppermint	Leaves
Marigold, French	Flowers
Onions, Brown	Skins
Privet	Leaves
Scaley buttons	Flowers
Willow peppermint gum	Leaves

ORANGE

Common name	Plant part
Bixa	Seeds
Everlasting daisy	Flowers

STRING

Common name	Plant part
Allamanda	Flowers
Banksia, Hill	Flowers
Basil	Leaves
Blue gum	Leaves
Bottlebrush	Leaves
Brazilian cherry	Berries
Brigalow	Leaves
Burdekin plum	Fruit
Camphor laurel	Leaves
Canna	Leaves
Daisy tree, Mexican	Leaves
Dogwood	Leaves
Dracaena	Leaves
Elkhorn	Leaves
Geranium, Red	Flowers
Gum, Blue	Leaves
Hill banksia	Flowers
Lasiandra	Flowers
Mexican daisy tree	Leaves
Oak	Acorns
Pine tree	Needles & cones
Polygala	Flowers
Potato vine	Berries
Queensland nut	Leaves
Rose	Petals & leaves
Sweet pea	Flowers
Verbena, Veined	Flowers
Verbena	Leaves
Wattle, White sallow	Leaves
Westringia	Leaves

SCARLET

Common name
Cochineal beetles

Part
Powder

RED

Common name
Argyle apple

Plant part
Leaves

PINK

Common name
Cochineal beetles

Part
Powder & liquid

WATERMELON

Common name
Cochineal beetles

Part
Liquid

PLUM

Common name
Lichen

Plant part
Whole

PURPLE

Common name
Blackberry nightshade
Corky passion vine
Deadly nightshade
Mulberry

Plant part
Berries
Berries
Berries
Berries

LAVENDER

Common name
Corky passion vine
Lichen

Plant part
Berries
Whole

Grevillea sp.

Pawpaw

2 Plants for Colours List

Common name	Botanical name	Plant part	Alum mordant
Agapanthus	Agapanthus umbellatus	Flowers	Greenish cream
Allamanda	Allamanda cathartica	Flowers	Golden fawn
Aluminium plant	Pilea cadierei nana	Leaves	Creamy gold
Ardisia	Ardisia crispa	Berries	Soft light grey
Argyle apple	Eucalyptus cinerea	Leaves	Rust
Asparagus fern	Asparagus sprengeri	Leaves	Pale lime green
Azalea	Rhododendron	Pink flowers	Pale yellow
Banksia, Hill	Banksia collina	Flowers	String
Banksia, Hill	Banksia collina	Leaves	Lemon
Basil	Ocimum basilicum	Leaves	Pale yellow
Beetroot	Beta vulgaris	Root	Soft gold
Bixa	Bixa orellana	Seeds	Soft gold
Blackberry nightshade	Solanum nigrum	Berries	Purple
Bleeding heart	Dicentra spectabilis	Flowers	Bright yellow
Bloodwood, Swamp gum	Eucalyptus ptychocarpa	Leaves	Yellow fawn
Blue gum	Eucalyptus tereticornis	Leaves	Yellow
Blueberry	Vaccinium corymbosum	Fruit	Purple/grey
Bottlebrush	Callistemon viminalis	Leaves	Soft gold
Bottlebrush	Callistemon 'Captain Cook'	Leaves	Gold
Bougainvillea	Bougainvillea buttiana	Flowers & leaves	Acid green
Bracken fern	Pteridium esculentum	Leaves	Soft green
Brazilian cherry	Eugenia brasiliensis	Berries	Fawn
Brigalow	Acacia harpophylla	Leaves	
Buderim soil	Red volcanic soil	Soil	Apricot
Burdekin plum	Pleioynium timorense	Fruit	String
Camellia	Camellia japonica	Dark red flowers	Fawn
Camphor laurel	Cinnamomum camphora	Wood shavings	Cream
Camphor laurel	Cinnamomum camphora	Leaves	Stringy gold
Canna	Canna indica	Leaves	Fawny brown
Carnations	Caryophyllus	Flowers	Golden fawn
Cassia	Cassia corymbosa	Leaves	Yellow green
Cassia	Cassia corymbosa	Flowers	Dull gold
Celery	Apium graveolens	Leaves	Creamy green
Chrysanthemum	Chrysanthemum maximum	Flowers	Yellow
Chrysanthemum	Chrysanthemum maximum	Leaves	Yellow
Clerodendron	Clerodendron speciosissima	Flowers	Creamy green
Cobbler's pegs	Bidens pilosa	Seeds	Apricot
Cochineal beetles	Dactylopius coccus	Powder	
Cochineal beetles	Dactylopius coccus	Liquid	
Coffee tree	Coffea arabica	Leaves	Gold
Coffee tree	Coffea arabica	Berries	Fawn
Comfrey	Symphutum officinale	Leaves	Cream
Coral tree	Erythrina	Leaves	Lime yellow
Coral tree	Erythrina	Flowers	Grey
Coral vine	Antigonon leptopus	Flowers	Deep gold brown
Corky passion vine	Passiflora suberosa	Berries	Purple
Cotoneaster	Cotoneaster	Leaves	Yellow fawn
Cypress, Monterey	Cupressus macrocarpa	Leaves	Yellow fawn
Dahlia	Dahlia pinnata	Flowers	Golden yellow
Daisy tree, Mexican	Montanoa bipinnatifida	Leaves	Cream

Chrome mordant	Copper mordant	Iron mordant	Unmordanted	Tin added
Pale olive	Brownish fawn	Fawn	Cream	
			String	
Soft gold	Light brown	Grey	Pinky fawn	
Golden grey	Soft brown grey	Light steel grey	Pale grey	
Rust	Deep rust	Brown red	Apricot to rust	
Goldy green	Greyish green	Grey green	Pale green	
Yellow green	Yellow green	Creamy grey	Cream	
Olive	Light fawn	Grey	String	
Citrus yellow	Yellow green	Soft brown	Cream	
Pale olive	Deep string	Grey	Pale cream	
Soft gold	Gold	Soft brown	Orange	
			Alum+Added Lime=Grey	
Golden yellow	Greenish yellow	Olive	Pale yellow	
Soft pale olive	Olive green	Grey	Deep string	String
Goldy grey	Greenish grey	Grey		
Gold brown	Brown	Grey	String	
Dull gold	Olive green	Dark brown		String
Citrus green	Pale green	Olive green		
Yellow green	Soft olive green	Pale olive green	Pale green	
Goldy fawn	Deep fawn	Soft brown	String	
			String	Greenish string
Fawn	Greenish fawn		Apricot	
Greenish fawn	Greenish fawn	Grey	Fawn	
Pale grey	Pale green	Pale gold		
Light brown	Olive gold	Grey brown	Pale string	
Brown	Deep brown	Dark brown	String	
Fawn	Fawn	Deep fawn	Fawn	
Citrus green	Green	Grey green	Pale green	
	Olive gold			
Lime green	Deep cream green	Olive green	Pale green	
Gold	Greenish yellow	Dark olive green	Pale yellow	
Gold	Greenish yellow	Dark olive green	Cream	
Light green	Pale green	Light grey	Cream	
Tan	Rust brown	Deep green		
				Pink to scarlet
				Pink-watermelon
Greenish gold	Deep brown	Black	Fawn	
Greenish brown	Soft brown	Brown	Fawn	
Citrus cream	Greenish cream	Grey	Cream	
Goldy green	Yellow green	Green grey	Pale green	
Gold grey	Grey	Grey	Pinky fawn	
				Gold
	Greenish purple	Greyish purple	Lavender	
Deep yellow fawn	Greenish fawn	Brownish fawn	Fawn	
Goldy fawn	Deep fawn	Grey fawn	Fawn	
Pale lemon	Soft green	Deep string	Cream	

Common name	Botanical name	Plant part	Alum mordant
Deadly nightshade	*Solanum nigrum*	Berries	Purple
Dogwood	*Cornus florida*	Leaves	Fawn
Dracaena	*Dracaena sanderiana*	Leaves	Creamy fawn
Elkhorn	*Platycerium bifurcatum*	Leaves	String
Everlasting daisy	*Helichrysum*	Flowers	Orange
Fennel	*Foeniculum vulgare*	Flowers	Bright yellow
Fennel	*Foeniculum vulgare*	Leaves	Soft pale green
Fishbone fern	*Nephrolepis bostoniensis*	Leaves	Green
Flame tree	*Brachychiton acerifolium*	Flowers	Cream
French marigold	*Tagetes patula*	Flowers	Gold
Fungi	*Polystictus multizoned*	Whole plant	
Geranium, Red	*Pelargonium*	Flowers	Soft olive green
Grape	*Vitis vinifera*	Leaves	
Grass	*Couch*	Leaves	Pale green
Grey box	*Eucalyptus argillacea*	Leaves	Dark lime green
Gum, Lemon-scented	*Eucalyptus citriodora*	Leaves	Pale yellow
Gum, Willow peppermint	*Eucalyptus nicholii*	Leaves	Ginger
Gum, Blue	*Eucalyptus tereticornis*	Leaves	Yellow
Gum, Mealy stringybark	*Eucalyptus cinerea*	Leaves	Rust
Gum, Narrow-leaf red ironbark	*Eucalyptus crebra*	Leaves	Fawn
Gum, Argyle apple	*Eucalyptus cinerea*	Leaves	Rust
Hackberry, Chinese	*Celtis sinensis*	Leaves	Greenish cream
Hibiscus	*Hibiscus syriacus*	Leaves	Greenish cream
Hibiscus, Orange	*Hibiscus syriacus*	Flowers	Cream
Hill Banksia	*Banksia collina*	Leaves	Lemon
Hill Banksia	*Banksia collina*	Flowers	String
Ironbark, Narrow-leaf red	*Eucalyptus crebra*	Leaves	Fawn
Jacaranda	*Jacaranda mimosaefolia*	Flowers	Pale fawn
Jasmine	*Jasminum officinale*	Flowers	Pale green yellow
Jasmine	*Jasminum officinale*	Flowers & leaves	Lime yellow
Kalanoa		Leaves	Light gold
Lasiandra	*Tibouchina urvilleana*	Leaves	Bright yellow
Lasiandra	*Tibouchina urvilleana*	Flowers	Creamy yellow
Lemon-scented teatree	*Leptospermum petersonii*	Leaves	Golden fawn
Lemon-scented gum	*Eucalyptus citriodora*	Leaves	Pale yellow
Lettuce	*Lactuca sativa*	Leaves	Very pale green
Lichen, Old man's beard	*Lichen fruticose*	Whole	Golden fawn
Lichen	*Lichen umbilicaria*	Whole	
Loquat	*Eriobotrya japonica*	Leaves	Fawn
Mango	*Mangifera indica*	Leaves	Olive green
Marigold, French	*Tagetes patula*	Flowers	Gold
Mealy stringybark	*Eucalyptus cinerea*	Leaves	Rust
Mexican daisy tree	*Montanoa bipinnatifida*	Leaves	Cream
Mint	*Mentha*	Leaves	Soft olive green
Mistletoe	*Loranthaceae*	Leaves	Cream
Moses-in-a-boat/basket	*Rhoeo discolor*	Leaves	Cream
Mulberry	*Morus nigra*	Berries	Purple
Narrow-leaf red ironbark	*Eucalyptus crebra*	Leaves	Fawn
Nasturtium	*Tropaeolum majus*	Flowers	Lime green
Nasturtium	*Tropaeolum majus*	Leaves	Pale green
Oak	*Quercus*	Acorns	String
Old man's beard	*Lichen fruticose*	Whole	Golden fawn
Onions, Brown	*Allium cepa*	Skins	Ginger

Chrome mordant	Copper mordant	Iron mordant	Unmordanted	Tin added
			Alum+Added Lime=Grey	
String	Greenish fawn	Grey	Fawn	Yellow
			String	
Greenish string	Pale green	Grey	Cream	
	Brown	Grey	Pale gold	
Golden yellow	Pale yellow	Pale olive green	Cream	
Yellow green	Soft grey gren	Olive green		
Soft leaf green	Leaf green	Olive green		
Creamy gold	Greenish gold	Pale grey	Pinky fawn	
Ginger	Deep olive	Brown	Cream	
	Cream			Pale cream
Yellow green	Soft olive green	Grey brown	String	
			Deep cream	
Gold green	Pretty green	Grey green	Pale green	
Goldy green	Olive green	Dark grey	Greenish cream	Greenish cream
Citrus yellow	Olive green	Grey brown	Cream	Golden fawn
Ginger	Greenish ginger	Brownish black	Ginger	
Soft pale olive	Olive green	Grey	Deep string	String
Rust	Deep rust	Brown red	Apricot to rust	
Golden green	Grey green	Light grey	Fawn	
Rust	Deep rust	Brown red	Apricot to rust	
Light green	Green	Green		
Pale olive	Greenish fawn	Pale grey	Pale cream	
Deep cream	Pale green			
Citrus yellow	Yellow/green	Soft brown	Cream	
Olive	Light fawn	Grey	String	
Golden green	Grey green	Light grey	Fawn	
Pale olive green	Greenish fawn	Greyish fawn	Fawn	Pale fawn
Leaf green	Leaf green	Greenish grey	Pale light green	
Goldy yellow	Olive green	Olive grey green	Greenish cream	
Deep gold	Gold	Brownish gold	Light gold	Light gold
Greenish yellow	Creamy green	Creamy olive	Greenish cream	
Pale olive	String	Grey	Cream	
Golden olive	Greenish gold	Soft brown	Fawn	
Citrus yellow	Olive green	Grey brown	Cream	Golden fawn
			Pale cream green	
Gold	Greenish fawn	Brown	Cream	
			Lavender to plum	
Deep fawn	Soft brown	Deep mid brown	Pinky fawn	
Ginger	Golden brown	Deep olive	Cream	
Rust	Deep rust	Brown red	Apricot to rust	
Pale lemon	Soft green	Deep string	Cream	
	Fawn	Greenish brown		
Greyish cream	Greenish grey	Pale green	Cream	
Greenish purple	Greenish purple	Grey purple	Grey purple	Good purple
Gold green	Grey green	Light grey	Fawn	
Citrus green	Soft green	Soft green	Greenish cream	
Citrus green	Soft green	Olive green	Soft green	
Golden fawn	Olive	Dark brown	Fawn	
Gold	Greenish fawn	Brown	Cream	
Browny ginger	Ginger	Browny ginger	Ginger	Yellow

Common name	Botanical name	Plant part	Alum mordant
Parsley	Petroselinum crispum	Leaves	Green, good
Passionfruit	Passiflora edulis	Skins	Cream
Passionflower	Passiflora	Leaves	Golden yellow
Pawpaw	Carica papaya	Leaves	Citrus yellow
Peppermint gum, Black	Eucalyptus nicholii	Leaves	Deep apricot
Pine tree	Pinus radiata	Needles	Fawn
Pine tree	Pinus radiata	Cones	Pale grey
Plumbago/leadwort	Plumbago capensis	Flowers & leaves	Greenish gold
Polygala	Polygala myrtifolia grandis	Flowers	Cream
Pomegranate	Punica granatum	Flowers	Deep gold
Potato vine	Solanum seaforthianum	Berries	Apricot
Potatoes, Red sweet	Impomoea batatas	Root	Cream
Prickly pear	Cactus	Fruit	Gold
Privet	Ligustrum vulgare	Leaves	Yellow
Purple King beans	Phaseolus vulgaris	Roots	Pale cream
Purple King beans	Phaseolus vulgaris	Stems	Yellow
Purple King beans	Phaseolus vulgaris	Seeds	Greenish cream
Purple cabbage	Brassica oleracea	Leaves	Pinky grey
Queensland nut	Macadamia	Leaves	
Red salvia	Salvia splendens	Flowers	Pinky fawn
Rosemary, Coastal	Westringia fruticosa	Leaves & flowers	Lime yellow
Rose	Rosa	Petals	Pale lime green
Rose	Rosa	Leaves	Light lime green
Sandalwood	Santalum	Leaves	Lime green
Scaley buttons	Leptorhynchus squamatus	Flowers	Gold
Scaley buttons	Leptorhynchus squamatus	Leaves	Light green
Scarlet sage	Salvia splendens	Flowers	Pinky fawn
Silky oak	Grevillea robusta	Leaves	Greeny gold
Spider plant	Chlorophytum comosum pictata	Leaves	Creamy fawn
Swamp bloodwood gum	Eucalyptus ptychocarpa	Leaves	Yellow fawn
Sweet pea	Lathyrus	Flowers & leaves	Pale green
Sweet pea	Lathyrus	Flowers	String green
Sword fern	Nephrolepis cordifolia	Leaves	Creamy fawn
Tallowwood	Eucalyptus microcorys	Wood chips	Gold
Tallowwood	Eucalyptus microcorys	Leaves	Light apricot
Teatree, Lemon-scented	Leptospermum petersonii	Leaves	Golden fawn
Tree fern, Rough	Cyathea australis	Leaves	
Turmeric	Circuma longa	Root	
Verbena, Veined	Verbena	Flowers	Light gold green
Verbena	Verbena	Leaves	Light gold green
Vitex	Vitex purpurea trifolia	Leaves	Soft grey green
Wandering Jew	Commelina cyanea	Leaves	
Wattle, White sallow	Acacia floribunda	Leaves	Yellow
Wattle, Brisbane	Acacia fimbriata	Leaves	Gold
Wattle, Zig-zag, Flat, Clay	Acacia macradenia	Leaves	Gold
Wattle, Silver	Acacia dealbata	Leaves	Deep fawn
Wetringia	Westringia fruticosa	Leaves	Yellow string
Westringia	Westringia fruticosa	Flowers & leaves	Pale green yellow
White cedar	Melia azedarach	Berries	Cream
Willow peppermint gum	Eucalyptus nicholii	Leaves	Ginger
Yesterday, today, tomorrow	Brunfelsia pauciflora	Flowers	Pale cream green

Chrome mordant	Copper mordant	Iron mordant	Unmordanted	Tin added
Pretty green	Pretty green	Pretty green	Pale green	
Pale green	Greenish grey			
Yellow green	Olive green	Dark olive green	Pale green	
Yellow green	Leaf green	Grey	Pale cream	
Rust	Olive rust	Brown black		
Fawn	Greenish fawn	Light brown	String	
Deep string	Greenish grey	Deep grey	Grey	
Golden brown	Chocolate brown	Greenish grey	Greenish fawn	
Yellow cream	Greenish cream	String	Cream	
Deep gold	Deep gold	Dark brown		Deep gold
Apricot fawn	String	Apricot fawn	Pale apricot	
	Greenish cream	Soft grey		
Olive green	Pale green			Ginger
Cream	Greenish cream	Cream	Cream	
Yellow	Greenish yellow	Pale green	Pale cream	
Pale olive	Fawn	Pale grey	Pale cream	
	Soft brown	Pinky grey	Pale grey	
			String	
Golden fawn	Greenish fawn	Pale grey	Grey fawn	
Golden yellow	Olive green	Grey black	Pale yellow	
Deep lime green	Olive green	Soft brown	String	Gold
Deep lime green	Olive green	Soft dark brown	Creamy string	
Golden green	Light olive green	Grey green	Greenish cream	
Ginger	Brownish gold	Black	Greenish gold	
Golden green	Olive green	Dark grey	Pale gold	
Golden fawn	Greenish fawn	Pale grey	Grey fawn	
Creamy green	Greeny grey	Soft grey	Cream	
Pale green	Pale green	Stringy green	Fawn	
Goldy green	Deep fawn	Dark string	Fawn	
Yellow green	Green	Grey green	Creamy fawn	
Olive green	Dark olive green	Dark brown	Fawny brown	
Gold	Olive brown	Dark brown	Apricot	
Golden olive	Greenish gold	Soft brown	Fawn	
	Leaf green			
	Green	Green		Golden yellow
Light gold	Goldy green	Greenish grey	String	
Light gold	Goldy green	Greenish grey	String	
	Light bright green			
Deep yellow	String	Pale grey	Cream	
Gold brown	Chocolate brown	Grey brown	Deep fawn	
Gold	Chocolate brown	Grey brown	Fawn	
Goldy brown	Greenish brown	Grey brown	Deep fawn	
Gold	Greenish gold	Grey	String	
Goldy olive	Olive green	Greenish grey	Greenish cream	
Greenish cream	Greenish cream	Cream		
Ginger	Greenish ginger	Brownish black	Ginger	
Yellow green	Pale olive	Grey green		

3 Botanical to Common Names List

Botanical name	Common name
Acacia dealbata	Wattle, Silver
Acacia fimbriata	Wattle, Brisbane
Acacia floribunda	Wattle, White sallow
Acacia macradenia	Wattle, Zig-zag, Flat, Clay
Acacia harpophylla	Brigalow
Agapanthus umbellatus	Agapanthus
Allamanda cathartica	Allamanda
Allium cepa	Onions, Brown
Antigonon leptopus	Coral vine
Apium graveolens	Celery
Ardisia crispa	Ardisia
Asparagus sprengeri	Asparagus fern
Banksia collina	Hill banksia
Beta vulgaris	Beetroot
Bidens pilosa	Cobbler's pegs
Bixa orellana	Bixa
Bougainvillea buttiana	Bougainvillea
Brachychiton acerifolium	Flame tree
Brassica oleracea	Purple cabbage
Brunfelsia pauciflora	Yesterday, today, tomorrow plant
Cactus	Prickly pear
Callistemon 'Captain Cook'	Bottlebrush
Callistemon viminalis	Bottlebrush
Camellia japonica	Camellia
Canna indica	Canna
Carica papaya	Pawpaw
Caryophyllus	Carnations
Cassia corymbosa	Cassia
Celtis sinensis	Hackberry, Chinese
Chlorophytum comosum pictata	Spider plant
Chrysanthemum maximum	Chrysanthemum
Cinnamomum camphora	Camphor laurel
Circuma longa	Turmeric
Clerodendron speciosissima	Clerodendron
Coffea arabica	Coffee tree
Commelina cyanea	Wandering Jew
Cornus florida	Dogwood
Cotoneaster	Cotoneaster
Couch	Grass
Cupressus macrocarpa	Cypress monterey
Cyathea australis	Tree fern, Rough
Dactylopius coccus	Cochineal beetle
Dahlia pinnata	Dahlia
Decentra spectabilis	Bleeding heart
Dracaena sanderiana	Dracaena
Eriobotrya japonica	Loquat
Erythrina	Coral tree
Eucalyptus argillacea	Grey box
Eucalyptus cinerea	Mealy stringybark
Eucalyptus citriodora	Lemon-scented gum
Eucalyptus crebra	Narrow-leaf red ironbark
Eucalyptus microcorys	Tallowwood
Eucalyptus nicholii	Willow peppermint gum
Eucalyptus ptychocarpa	Swamp bloodwood gum
Eucalyptus tereticornis	Gum, Blue
Eugenia brasiliensis	Brazilian cherry
Foeniculum vulgare	Fennel
Grevillea robusta	Silky oak
Helichrysum	Everlasting daisy
Hibiscus syriacus	Hibiscus, Orange
Impomoea batatas	Potatoes, Red sweet
Jacaranda mimosaefolia	Jacaranda
Jasminum officinale	Jasmine
Lactuca sativa	Lettuce
Lathyrus	Sweet pea
Leptorhynchus squamatus	Scaley buttons
Leptospermum petersonii	Teatree, Lemon-scented
Lichen fruticose	Old man's beard
Lichen umbilicaria	Lichen
Ligustrum vulgare	Privet
Loranthaceae	Mistletoe
Macadamia	Queensland nut
Mangifera indica	Mango
Melia azedarach	White cedar
Mentha	Mint
Montanoa bipinnatifida	Mexican daisy tree
Morus nigra	Mulberry
Nephrolepis bostoniensis	Fishbone fern
Nephrolepis cordifolia	Sword fern
Ocimum basilicum	Basil
Passiflora	Passion flower
Passiflora edulis	Passionfruit
Passiflora suberosa	Corky passion vine
Pelargonium	Geranium, Red
Petroselinum crispum	Parsley
Phaseolus vulgaris	Purple King beans
Pilea cadierei nana	Aluminium plant
Pinus radiata	Pine tree
Platycerium bifurcatum	Elkhorn
Pleioynium timorense	Burdekin plum
Plumbago capensis	Leadwort, Plumbago
Polygala myrtifolia grandis	Polygala
Polystictus multizoned	Fungi
Pteridium esculentum	Bracken fren
Punica granatum	Pomegranate
Quercus	Oak
Red volcanic soil	Buderim soil
Rhododendron	Azalea

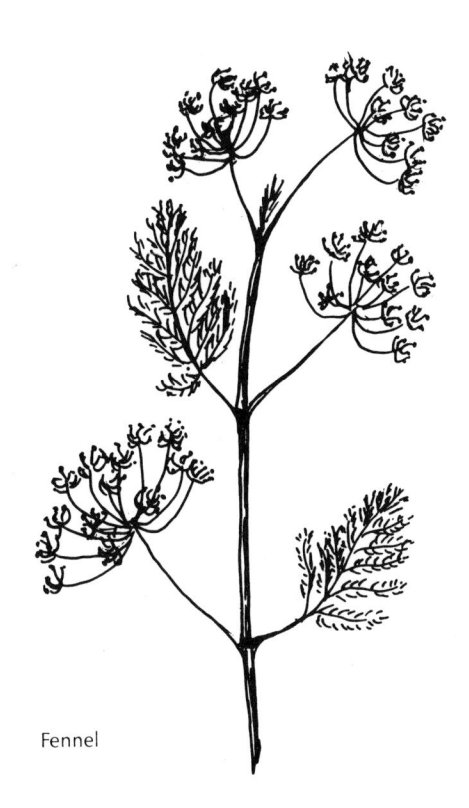

Rhoeo discolor	Moses-in-a-boat/basket	*Tibouchina urvilleana*	Lasiandra
Rosa	Rose	*Tropaeolum majus*	Nasturtium
Salvia splendens	Scarlet sage	*Vaccinium corymbosum*	Blueberry
Santalum	Sandalwood	*Verbena*	Verbena
Solanum nigrum	Blackberry/deadly nightshade	*Vitex purpurea trifolia*	Vitex
Solanum seaforthianum	Potato vine	*Vitis vinifera*	Grape
Symphutum officinale	Comfrey	*Westringia fruticosa*	Westringia
Tagetes patula	Marigold, French		

Fennel

4 Wood Shavings for Colours List

Common name	Botanical name	Alum mordant	Chrome mordant
Australian native cherry	*Exocarpus cupressiformis*	Light fawn	Mid fawn
Black bean	*Castanospermum*	Yellowish fawn	Brownish fawn
Black wattle	*Callicoma serratifolia*	Ginger brown	Light brown
Brazilwood	*Caesalpinia echinata*	Coppery fawn	Brown
Camphor Laurel	*Cinnamomum camphora*	Golden yellow	Deep gold yellow
Cypress pine	*Callitris cupressiformis*	Pale lemon	Deep lemon
Ebony	*Diospyros ebenum*	Deep brown	Deep brown
Hornbeam ironwood	*Carpinus betulus*	Rich gold	Light gold
Iron bark	*Eucalyptus racemosa*	Golden yellow	Deep golden yellow
Merbue	*Intsia palembanica*	Rich gold	Brownish gold
Mulberry (green timer)	*Morus nigra*	Gold	Golden brown
Obeche	*Triplochiton*	Cream	Cream
Padauk	*Pterocarpus*	Rust red	Deep soft red
Purple heart	*Peltogyne*	Mid green	Leaf green
Queensland maple	*Flindersia brayleyana*	Golden yellow	Ginger fawn
Red cedar (green)	*Toona australis*	Soft red	Pinky red
Red cedar (seasoned)	*Toona australis*	Light brown	Soft brown
Red river gum	*Eucalyptus camaldulensis*	Golden yellow	Deep golden yellow
Red siris	*Albizia toona*	Gold	Deep gold
Rose alder	*Caldcluvia australiensis*	Creamy string	String
Rosewood	*Dysoxylum fraseranum*	Pinky deep fawn	Soft brown
Tallow wood	*Eucalyptus microcorys*	Fawny green	Pale green
Tasmanian blackwood	*Acacia melanoxylon*	Ginger brown	Brown
Wenge	*Millettia laurentii*	Ginger chocolate brown	Chocolate brown
Yellow siris	*Albizia xanthoxylon*	Yellow	Deep yellow

Camphor Laurel

Copper mordant	Iron mordant	Unmordanted
Deep fawn	Brownish fawn	Fawn
Brownish fawn	Light brown	Fawn
Greenish brown	Soft brown	Deep fawn
Brown	Coppery brown	Copper
Golden brown	Olive green	String
Fawn	Fawn	Pale fawn
Chocolate brown	Chocolate brown	Soft brown
Brownish gold	Greenish fawn	Fawn
Deep fawn	Greenish fawn	Fawn
Greenish brown	Rich brown	Golden yellow
Soft brown	Greenish brown	Pale gold
Brownish cream	Silver grey	Cream
Deep soft red	Plum red	Soft red
Emerald green	Grey green	Soft green
Soft brown	Soft grey	Fawn
Deep pinky red	Grape	Pinky fawn
Soft brown	Brown	Fawn
Golden brown	Light brown	Golden yellow
Deep gold	Pale brown	Pale gold
String	Silver grey	Cream
Soft brown	Soft light brown	Fawn
Mid green	Olive green	Greenish fawn
Brown	Olive brown	Yellow fawn
Chocolate brown	Chocolate brown	Light brown
Deep string	Pale grey	Fawn

References

Dunn, Margaret Hanlon, and Barbara O'Leary: *Colour Me A Garden*, Golden Press, Sydney, 1987

Greig, Denise: *The Australian Gardener's Wildflower Catalogue*, Angus & Robertson, Sydney, 1987

Harris, Thistle Y.: *Wildflowers of Australia*, (7th edition), Angus & Robertson, Sydney, 1973

Holliday, Ivan, and Ron Hill: *A Field Guide to Australian Trees*, Rigby, Sydney, 1969

Holliday, Ivan, and Geoffrey Watton: *A Gardener's Guide to Eucalypts*, Rigby, Sydney, 1980

Lamp, Charles, and Frank Collet: *A Field Guide to Weeds in Australia*, (revised edition), Inkata, 1979

Red River Gum

Index

Acrylic fibres, 10
Alum, 12, 19, 21
Aluminium plant, 12
Ammonia, 13, 15, 27
Ancient dyes, 7
Animal fibres, 7
Ardisia, 17
Argyle apple, 18

Bags, 8, 42
Bargello, 48
Bark, 8, 13, 15, 42
Beetroot, 48
Berries, 7, 8, 13, 15, 39, 40, 42-3
Bixa, 17
Black bean, 36
Bleach, 27, 39
Blue, 7, 45
Botanical names, 24, 60
Bougainvillea, 17
Bracken, 8, 38

Cactus, 7, 26
Camellia, 12
Camphor laurel, 62
Cassia, 17, 42
Celery, 12, 17
Chemicals, 15, 19, 21-2, 38
Children's dyeing, 38
Children-precautions, 23
Chrome, 12, 19, 21, 23, 45
Chrysanthemums, 12
Cleaning yarn, 11, 24
Cochineal, 7, 26
Collecting, 8, 13, 27, 39-42
Colour lists, 45-59, 62-3
Common names, 60-1
Cones, 15
Copper, 12, 19, 21, 23, 45
Copper sulphate, 19, 21
Cotton, 30
Cream of tartar, 21-2

Drying, 8, 10
Dyeing method, 24
Dye experiment sheet, 9
Dye making, 12
Dye pots, 12, 14

Eucalypts, 12, 22, 31, 38
Eucalyptus cinerea, 18, 38
Eucalyptus nicholii, 24
Exhaust, 14, 27
Experiment sheet, 9
Extraction times, 15

Fading, 7, 13, 27
Fennel, 61
Ferrous sulphate, 19, 22, 45
Fibres, 7, 10
First aid, 23
Flame tree, 31
Fleece, 6, 10, 24, 65
Flowers, 8, 13, 17, 30-2, 34, 38-40, 42
Fruit, 13, 34

Grasses, 8
Grevillea, 47, 53

Heating, 16, 39
Herbs, 13, 39
Hints, 39
Hotplate, 16

Indigofera, 7
Insect dye, 7
Iron, 12, 19, 22
Ironbark, 30

Jacaranda, 31, 44

Knotting system, 19

Lichen, 27, 32, 39
Loquat leaves, 12

Marigolds, 16, 33, 43
Marking system, 19
Matted wool, 10, 19, 24
Mixing colours, 39
Mordanting, 19
Mordants, 12, 19, 21-3, 27
    disposing of residues, 22
Mulberries, 38
Murex, 7

Nasturtium, 12, 38
Non-reactive pots, 12, 14
Nuts, 13

Odours, 16, 43-4
Onion skins, 12, 16, 33, 38

Passiflora, 14, 34
Passionfruit skins, 46
Pawpaw, 34, 53
Pine cones, 15
Plant fibres, 7
Plant list, 60
Plumbago, 34
Poisoning, 23, 38, 39
Potassium aluminum sulphate, 19
Potassium dichromate, 19
Pots, 12, 14
Pressing plants, 40
Projects, 12-13, 38-9
Pruning, 13
Purple, 7, 38, 45

Quantities, 14
Queensland maple, 29

Rainwater, 15
Ratios, 14
Records, keeping of, 9
Red, 7, 21, 26-7, 45
Red cedar, 30
Red river gum, 63
Roots, 13

Roses, 12, 34, 43
Rosewood, 34

Sample book, 40
Samples, 38-9, 42
Saucepans, 12, 14
Scarlet dye, 26
Seeds, 13, 17
Shrinkage, 10
Silk, 36-7
Silky oak, 47
Simmering, 10, 12, 14, 19, 24
Skeins, 12, 19, 39
Soap powder, 10
Soil, 43-4
Stannous chloride, 19, 22, 26
Storage, 42
Stories, 43
Striped square, 39

Tasmanian blackwood, 29
Tea-tree, 32
Testing, 15, 24, 27, 39
Times for dyeing, 15
Tin, 12, 19, 22, 26
Tree ferns, 47
Tyrian purple, 7

Vegetables, 13, 17, 33
Vinegar, 27
Vitex, 47

Washing wool, 10
Water, 12, 14-15
Wattle, 48
Weather, 8, 13
Weeds, 8
Weighing, 10, 14, 24
Westringia, 39, 47
Wildflowers, 8
Wood shavings, 28-9, 62-3
Wool, 6-7, 10, 14-15, 19, 36

Yarn, 10, 14